TQM in Action

JOIN US ON THE INTERNET VIA WWW, GOPHER, FTP OR EMAIL:

WWW: http://www.thomson.com
GOPHER: gopher.thomson.com
FTP: ftp.thomson.com
EMAIL: findit@kiosk.thomson.com

A service of I(T)P®

TQM in Action

A PRACTICAL APPROACH TO CONTINUOUS PERFORMANCE IMPROVEMENT

Second edition

John Pike
Professor of Organization Development, and Director of the Total Quality and Innovation Management Centre, Anglia Business School, Anglia Polytechnic University, UK.

and

Richard Barnes
Barnes Associates, Consultants in Quality and Human Resource Management, and Principal Associate Consultant, Total Quality and Innovation Management Centre

CHAPMAN & HALL
London · Weinheim · New York · Tokyo · Melbourne · Madras

Published by Chapman & Hall, 2–6 Boundary Row, London SE1 8HN, UK

Chapman & Hall, 2-6 Boundary Row, London SE1 8HN, UK

Chapman & Hall GmbH, Pappelallee 3, 69469 Weinheim, Germany

Chapman & Hall USA, 115 Fifth Avenue, New York NY 10003, USA

Chapman & Hall Japan, ITP-Japan, Kyowa Building, 3F, 2–2–1
Hirakawacho, Chiyoda-ku, Tokyo 102, Japan

Chapman & Hall Australia, 102 Dodds Street, South Melbourne, Victoria
3205, Australia

Chapman & Hall India, R. Seshadri, 32 Second Main Road, CIT East,
Madras 600 035, India

First edition 1994

Second edition 1996

Reprinted 1996

© 1994, 1996 John Pike and Richard Barnes

Typeset in 10/12 Times by Mews Photosetting, Beckenham, Kent
Printed in Great Britain by Page Bros (Norwich) Ltd

ISBN 0 412 71530 9

Apart from any fair dealing for the purposes of research or private study, or criticism or review, as permitted under the UK Copyright Designs and Patents Act, 1988, this publication may not be reproduced, stored, or transmitted, in any form or by any means, without the prior permission in writing of the publishers, or in the case of reprographic reproduction only in accordance with the terms of the licences issued by the Copyright Licensing Agency in the UK, or in accordance with the terms of licences issued by the appropriate Reproduction Rights Organization outside the UK. Enquiries concerning reproduction outside the terms stated here should be sent to the publishers at the London address printed on this page.

The publisher makes no representation, express or implied, with regard to the accuracy of the information contained in this book and cannot accept any legal responsibility or liability for any errors or omissions that may be made.

A catalogue record for this book is available from the British Library

Library of Congress Catalog Card Number: 95-74652

∞ Printed on permanent acid-free text paper, manufactured in accordance with ANSI/NISO Z39.48-1992 and ANSI/NISO Z39.48-1984 (Permanence of Paper).

Contents

Preface

When we wrote the manuscript for the first edition of this book, Total Quality Management, as a concept, was still relatively new to many organizations. Today, only a few years on, it is unlikely that there are many organizations which have not heard about it. This does not mean, however, that because organizations have heard about it they are practising it. Indeed, from recent research that we have carried out only 53.7% of the respondents claimed to be practising Total Quality Management processes. Furthermore, we know from our consulting experiences that many organizations which claim to be practising Total Quality are only doing so in a partial, or dare we say it, sometimes 'half-baked' manner. Even companies which try hard to succeed do not find it easy and many will say that they are still in the learning situation several years on from the start of the implementation process.

Total Quality Management in the spirit of Continuous Improvement is itself moving on. Inevitably there will always be those who wish to introduce a new 'flavour of the month' or wish to give TQM a new label in order to appear to have the latest, up-to-date, fashionable approach to sell to unsuspecting clients. These 'same wine, different label' merchants only serve to confuse managers who find it difficult enough as it is to grasp the ever increasing amount of jargon that characterizes the field. We can dismiss these efforts with the contempt they deserve. What we can't dismiss, however, are the genuine developments and refinements to the processes of TQM which have grown out of the insights gained by companies whose experience might be termed 'best practice'. We have learned for example that 'measurement' plays a key role in successful implementations, and that companies are striving to devise appropriate measures suitable to their own circumstances to gauge their successes. We have known for a long time that successful approaches require the commitment of top management, but what many organizations have not foreseen is the incredible amount of training that is required for members of the steering group, for middle managers and for employees in general, in order to bring about cultural change, which lies at the heart of TQM.

In this new edition we have included a case study of how one company has set about implementing Total Quality processes throughout the organization over a period of several years. The case provides a considerable amount of detail to illustrate what is really involved in making it happen. In addition to describing what actually happened, we have also presented the reflections of key managers and some of the staff who were involved. Very rarely will practising managers be able to gain such detailed insights into TQM in action.

As we said in the Preface to the first edition, the key to a successful implementation strategy is to design it right in the first place. We hope this second edition will provide even more insights to help managers master this task.

Preface to the first edition

Since we began writing this book, the world in which we live has changed radically. The Single European Market has formally come into being; the break up of the Soviet Union and the Unification of Germany have continued to disturb the economic order of world markets. A new President has taken over in the United States and pledged to try to halt the decline of their domestic economy while the recession in Britain continues to decimate manufacturing industry. Meanwhile the need for new approaches to the way we respond to these pressures for change is beginning to dawn. The interest in the Total Quality Management approach is increasing and more and more organizations are now claiming to have adopted Total Quality Management processes.

Inside this book we make reference to many of these organizations. Some will have succeeded, and no doubt some will have fallen by the wayside in the time since we first put pen to paper. Total Quality Management is no guarantee of success, any more than a healthy lifestyle guarantees permanent good health – it simply increases your chances. It was our intention to write a practical guide for managers to assist them to implement a Total Quality Management approach suitable to their own circumstances. We believe the key to a successful implementation strategy is to design it right in the first place. We hope this book helps.

Dedicated to the memory of Frank Barnes, Richard's brother who died tragically in a hot air balloon accident during the writing of this second edition.

Acknowledgements

This book is the product of our many years of experience working in, or as consultants to, a wide spectrum of organizations; large and small – public and private, manufacturing- and service-based, both here in the United Kingdom and in other parts of the world. Over the years we have been privileged to work with managers at all levels and had the opportunity to share the experiences of learning with and from each other in pursuit of solving real problems facing these managers. We would like to thank all of them for helping us to understand how organizations work and for shaping our thinking about how best to assist them to manage the processes of change, development and continuous improvement.

We would also like to thank Professor John L. Davies, Dean of the Anglia Business School, Anglia Polytechnic University, for his encouragement and support and for creating the conditions in which our work has been able to grow. To our colleagues in the Total Quality Management Research and Development Centre we would like to express our appreciation. Special mention should be made of Rick Waterhouse upon whose ideas we have drawn freely, as well as David Wood, Chairman of our Advisory Board, and Bob Davies who, when at the Engineering Industry Training Board, helped us to get started as a Centre. No work of this kind would have been possible without the permission of publishers to reproduce material from other authors. We would like to thank them all, especially the TQM magazine, for allowing us to do so.

We are especially grateful to our spouses, Margaret and Jo, and our families, for their support, help and constructive criticism from a reader's perspective.

Finally we owe considerable thanks to Mrs Ann Williams, who typed some of the initial chapters, and to Mrs Lesley Parry, secretary of the Total Quality Management Research and Development Centre, whose patience, enthusiasm and willingness to work hard at the word processor, ensured that we met our deadlines.

PART ONE

The Need for TQM

1 Introduction

This book has several main purposes. We wish to demonstrate that there is a need for a radical reappraisal of traditional management practices if organizations, particularly in the United Kingdom and the United States, are to respond effectively to the competitive demands of the international marketplace of the future. We also wish to provide managers with new insights into a powerful approach, that of Total Quality Management or Managing for Continuous Improvement as it is sometimes called, which is already proving in many companies to be an effective strategy for meeting these competitive demands. It is our intention to review some of the most commonly adopted approaches to the implementation of Total Quality Management and to provide managers with some practical guidelines to enable them to design and implement their own Total Quality Management Strategy, customized to their own organization.

We have set out to provide managers with a rational model and menu of ideas and suggestions for making the Total Quality Management process work right first time.

The book is structured in two parts. Part One begins by posing the question 'Can manufacturing industry survive?' and sets out the context in which managers in British and American industry find themselves. The desperate trouble that we are in unquestionably demonstrates that something different must be done. Our traditional approach to managing our organizations especially in the field of quality management has failed. We cannot continue to bury our heads in the sand and obdurately refuse to recognize that anything is wrong. We therefore point out some of the weaknesses of the traditional approach before getting down to the business of examining what an alternative approach, Total Quality Management, is all about. Many organizations have already embarked on the process of installing Total Quality Management into their organizations. Unfortunately evidence is beginning to emerge that the results are not as desirable as expected. Why? What have they done wrong? Is it the implementation process that they have adopted, the philosophy itself or some other factors which are getting in the way? We believe that there are 'many ways of skinning a cat'. Different organizations faced with different problems, different cultures, need different approaches from each other to bring about fundamental improvements.

Part Two of the book sets out to provide managers at any stage of the implementation process with answers to such questions as:

What do I have to do?
How do I do it in practice?
What do I do next?

Before we get to these practical considerations let us take a few steps back from what we need to do in our individual companies and take a look at the big picture. Why do we need to do anything?

1.1 CAN MANUFACTURING INDUSTRY SURVIVE?

Many years ago one of us, on taking up a new job, asked his boss 'What are our key objectives? 'SURVIVAL' came back the laconic reply. At that time he was really talking about his own personal survival within the company. Unfortunately he didn't achieve his objective. Today, many chief executives' and managing directors' minds are focused on the question of survival, not simply in terms of their own careers but for their companies as well. Such has been the ferocity of competition from overseas, combined with a hostile economic climate at home, that large numbers of companies have gone bankrupt and disappeared into oblivion. In both the United States and in Britain many traditional industries have faced extinction and companies which had become household names no longer trade. The USA was at one time the world's largest trading nation and leading economic power. Now its trade surpluses have shrunk and it has become the world's largest debtor country. In the United Kingdom the large balance of payments surplus of the early 1980s was transformed into a very large deficit by the end of the decade. Britain's total trading deficit with the rest of the world in 1990 was £14.6 billion.[1] During that period, described by some as 'the lost decade', British manufacturing industry had taken a hammering. As we entered the 1990s one report about longer term trends in British manufacturing industry noted that there was virtually no net investment in manufacturing industry during the 1980s and that it was conceivable that Britain would end up with no significant British owned manufacturing industry at all.[2]

Thus the question 'Can manufacturing industry survive'? is not an idle one for either American or British managers.

Is it too late? Are we too far behind to catch up with the competition? Do we need to do more than catch up – shouldn't we try to overtake? Is it not true that as fast as we get better, so too, do our competitors, thus widening the gap between us rather than narrowing it? If it is not too late, what is already being done to bring about improvements and what else needs to be done, not only to ensure survival but to promote growth as well? These are the big issues facing governments and the directors and senior managers of our wealth-creating industries today.

1.1.1 The scale of the problem

Many writers have drawn attention to the nature of the problems facing both the British and United States', economies. The catalogue of complaints includes:

- A lack of competitiveness of companies in global markets
- A fall in the growth of productivity
- Balance of Payments deficits in the USA and UK
- A decline in market share of world manufacturing exports alongside an increase in the share of imports
- A fall in real wages of manufacturing production workers
- Low levels of investment in manufacturing industry
- High interest rates

The question we must ask, however, is to what extent are these pessimistic assertions based on fact and how far are they merely opinions of the economic hypochondriacs determined to find something wrong with the health of our economies, regardless of the true situation. Let us look at some official statistics.

Gross Domestic Product 1960–1990

GDP 1960
In 1960 the value of Gross Domestic Product (GDP) at current prices and current exchange rates for the world top seven economies (in billions of US dollars) was as shown in Table 1.1.

GDP 1990
Thirty years later the relative position of each of these seven economies as measured by Gross Domestic Product at current prices and current exchange rates (in billions of US dollars) was as shown in Table 1.2.

What does this tell us? First of all it shows that the USA was still the world's largest economy, whilst West Germany and France had retained their respective positions of third and fourth largest economies in the league table. It also shows that the United Kingdom slipped from second place to sixth, Canada slipped from sixth place to seventh, and Italy rose from seventh place to fifth. Finally it shows that Japan rose from fifth place

Table 1.1 GDP for the world top seven economies in 1960

Position	Country	GDP ($ billion)
1	USA	513.38
2	UK	72.40
3	West Germany	72.07
4	France	60.90
5	Japan	44.47
6	Canada	40.41
7	Italy	39.73

(Source: OECD National Accounts Main Aggregates, Vol 1, 1960–1990, OECD Paris 1992)

Table 1.2 GDP for the world top seven economies in 1990

Position	Country	GDP ($ billion)
1	USA	5132.00
2	Japan	2942.93
3	West Germany	1488.21
4	France	1190.77
5	Italy	1090.75
6	United Kingdom	975.15
7	Canada	570.15

(Source: OECD National Accounts Main Aggregates, Vol 1, 1960–1990, OECD Paris 1992)

to second. The percentage change in GDP between 1960 and 1990 for the G7 countries is shown in Figure 1.1.

Here we can see that the USA grew at an average rate of three and a quarter percent a year, yet despite this, its share of industrialized output decreased from 57 per cent in 1960 to 33 per cent today. The chart also confirms that the Japanese rate of growth has been extremely rapid, averaging six and a half per cent a year since 1960.

Rate of growth 1960–1990

However, if position in the league table appears significant, what is clearly much more significant are the rates of growth of each of these countries.

From 1960 to 1990 the countries in the Group of Seven (G7) grew by the percentages shown in Table 1.3. In other words Japan grew throughout the period by more than six and one-half times the rate of the United States and nearly five times as fast as the United Kingdom. Thus the world's two largest economies in 1960 have exhibited, amongst the top seven, the most sluggish growth over the 30-year period. This surely must be a major cause for concern for governments and managers of both of these economies.

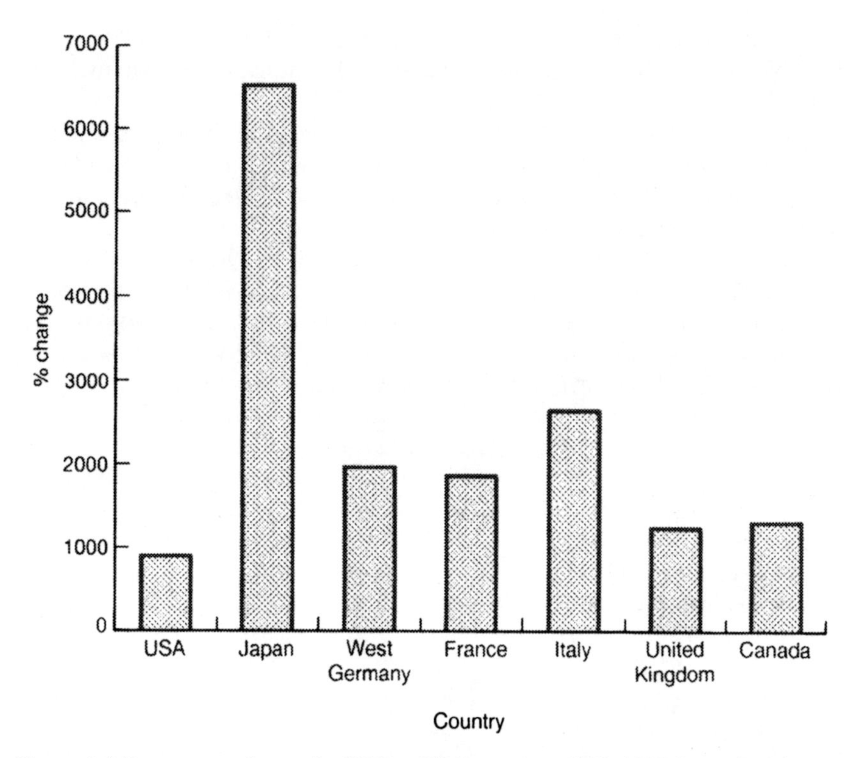

Figure 1.1 Percentage change in GDP – G7 Countries, 1960–1990 (compiled from OECD National Accounts 1960–1990, OECD, Paris).

Table 1.3 GDP rates of growth for the world top seven economies in period 1960–1990

Position	Country	Growth (%)
1	Japan	6617
2	Italy	2745
3	West Germany	2064
4	France	1955
5	Canada	1411
6	UK	1346
7	USA	999

Source: OECD Economic Survey for Sweden 1990–1991, basic statistics international comparisons – July 1992

GDP per capita

So far we have been looking at the size of the world's largest economies as measured by GDP in billions of US dollars. This is not the same thing, however, as GDP per capita, which takes account of the differences in the size of the population for each country. OECD figures for 1989 reveal GDP per capita (in rank order) measured in US dollars as shown in Table 1.4.

On this basis neither the USA nor the UK appear in the top seven. The USA with a figure of 20 639 is eighth and the UK with a figure of 14 642 is eighteenth overtaken by Iceland, Germany, Luxembourg, France, Australia, Austria, Belgium, Netherlands and Italy.

The difficulty, however, with these figures is that whilst they show the relative size of GDP per capita, it is still not a fair basis for comparison if we want to know how well off people are, because the figures do not take into account the differences in domestic costs. To allow for these differences the OECD has devised a means of making such comparisons, known as 'purchasing power parities' (PPPs). Purchasing power parities equalize the purchasing power of the different countries, so that, for example, £100 converted into US dollars at the PPP exchange rate would buy the same amount of a standard basket of goods and services in the USA as £100 does in the UK. Using these figures, OECD statistics reveal, for 1989, the figures shown in Table 1.5.

Here we can see that the USA tops the list. It is noticeable, however, that neither Germany nor the United Kingdom are in the top seven. Sweden with a figure of 15 511 is eighth, Finland with 15 030 is ninth, whilst West Germany with 14 985 is tenth and the United Kingdom with 14 345 is thirteenth in the league table.

1.1.2 The changing structure of the industrialized economies

The varying rates of growth of each of the countries listed above, conceals the way in which the structure of each of these economies has changed markedly over time. Taken as a whole, the output in the industrialized countries over the 30-year period has shifted considerably away from

Table 1.4 The top seven countries as measured by GDP per capita in 1989

Position	Country	GDP per capita ($)
1	Switzerland	26 350
2	Japan	23 305
3	Finland	23 370
4	Sweden	22 360
5	Norway	21 341
6	Canada	20 783
7	Denmark	20 685

Source: OECD Economic Survey for Sweden 1990–1991, basic statistics international comparisons

Table 1.5 The top seven countries as measured by GDP per capita in purchasing power parities in US dollars in 1989

Position	Country	GDP per capita ($, ppp based)
1	USA	20 629
2	Canada	19 305
3	Switzerland	17 699
4	Luxembourg	17 192
5	Norway	16 422
6	Iceland	15 870
7	Japan	15 712

Source: OECD Economic Survey for Sweden 1990–1991, basic statistics international comparisons

agriculture and manufacturing towards services. Figure 1.2 shows these changes.

Yet again these overall figures conceal wide differences within each of the industrialized countries. For example, agriculture accounts for 15% of output in Greece and Turkey but only 1¼% in the UK.

It is beyond our current scope to examine what has happened to manufacturing industry throughout each of the major industrialized countries over the last 30 years. However, to enable us to obtain a proper perspective on the problems facing both the United States and the United Kingdom in their attempts to face up to the competition coming from major competitors, such as Japan and West Germany, it is worth briefly highlighting some of the key features of manufacturing performance for these four economies.

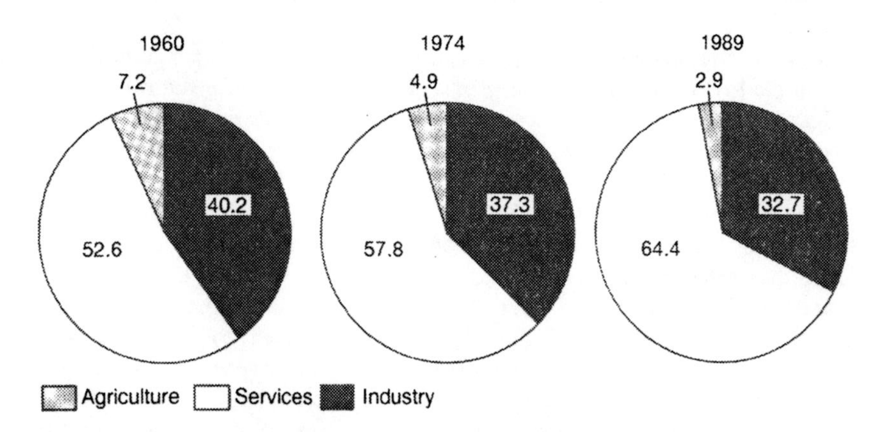

Figure 1.2 Percentage shares of output in the industrialized countries during period 1960–1990 (source: OECD Historical Statistics).

1.1.3 The decline of manufacturing industry

United Kingdom

There is no question that manufacturing industry in the United Kingdom has been in decline. In the 19th century the United Kingdom was the world's leading manufacturing nation but as other countries' manufacturing capacities grew, this supremacy was lost. The United Kingdom was overtaken by the USA in the early part of the 20th century and since then its manufacturing output has also been exceeded by Japan, Germany and France. Having lost its competitive edge, UK manufacturing industry's share of GDP shrank from 36.5% in 1960 to 23.5% in 1991. In other words it shrank by more than one-third or 35.6%. Over the same period the service sector grew from 45% to approximately 60%. This long-term decline has been exacerbated in recent years by the recession of 1980–1981 and also by the recession in the early 1990s. According to an Economist Intelligence Unit (EIU) report, the 1980–1981 recession effectively eliminated about 15% of UK manufacturing capacity[3]. This second recession, which is even more severe, has also affected manufacturing output and may eliminate at least an equal amount of capacity by the time it is over. The EIU report also points out that in 1983 the UK became a net importer of manufactured goods for the first time since the Industrial Revolution. The industries which were most severely affected, such as iron and steel, motor vehicles and textiles, were those which were most vulnerable to international competition. The implications of this reduction in the size of the industrial base is quite worrying. It is feared that the industrial base in the UK is now too small to meet the needs of the economy and any expansion will lead to an increase in imports and enlarge the already large trade deficit.

In these circumstances, and with the completion of the arrangements for the single European Market at the end of 1992, the need for manufacturing industry to become more competitive is paramount. In the Annual Economic Report of the Commission of European Communities for 1991–1992, it is noted that 'the share of UK GDP accounted for by investment has typically been lower than elsewhere in the community throughout the post-war period.' It also states that 'to secure a firm basis for higher medium term growth it will be essential to raise the investment share'[4]. To achieve this means a higher level of domestic saving and higher corporate savings if the alternative of higher capital inflows at higher rates of interest is to be avoided. Higher corporate savings will require higher profitability, which in turn can only be achieved through better working methods, lower costs and improvements in productivity levels.

United States

If there is no question about the decline of manufacturing industry in the United Kingdom, there appears to be a debate about what has happened in the United States. In a recent article in the *Harvard Business Review* (HBR) *Is America in Decline*? attention is drawn to a collection of books and articles about the American economy and its prospects, which highlight the arguments for and against the notion of decline[5]. Reference is made to an Office of Technology Assessment report called *Competing Economies: America, Europe and the Pacific Rim* and also to a report of the Competitiveness Policy Council entitled *Building a Competitive America*. Both reports, emerging from Washington, argue that America is in decline. The first argues that the USA's share of world manufacturing exports has declined in recent decades whilst its share of imports has risen. It also notes that the real wages of manufacturing production workers have fallen since the late 1970s.

The second, The Competitiveness Policy Council report, concludes that the United States' economic competitiveness 'is eroding slowly but steadily'. However, the author of the HBR article maintains that

> it is not clear that the USA has lost its competitive edge . . . Take manufacturing. The productivity of US manufacturing companies has improved noticeably since the mid 1980's when the competitiveness cry first arose. . . . In a whole set of industries some of which were prematurely given up for lost, US manufacturers are staging strong revivals and gaining strong export positions. Very simply, manufacturing no longer provides the unequivocal evidence of US decline that perhaps it once did.

The Economist Intelligence Unit's Country Profile for the USA for 1991–1992 recognizes that there have been structural changes in the US economy including the decline of 'smokestack' industries[6, p.29]. It also notes that because of the dollar's overvaluation between 1981 and 1985 there was a flood of imports which dented the dominance of US manufacturing industry in the domestic market. It was only after the depreciation of the US dollar after 1985 that US manufacturers got the chance to fight back and regain some of the ground they had lost in foreign markets.

The story of the American economy seems to be one of peaks and troughs but with an underlying trend of steady growth. This has not, however, been as fast as Japan, Germany or some other industrialized countries. Average growth for 1960–1970 was 3.8%, and from 1970–1980 it was 2.8%[6, p.13]. The Reagan boom boosted the average annual growth rate by increasing real GDP by 10% in the 2-year period 1983–1984, but the economy is now faced with a massive trade deficit of over 100 billion US dollars (1989)[7].

Japan

Manufacturing in Japan is central to the strength of its economy. As we have already seen, the success of the Japanese economy with its average annual growth rate of 6.5% has been one of the most spectacular features of post-war development of the industrial economies. It is now one of the richest countries in the world on a per capita basis and second only to the United States in absolute terms. It is also the world's largest creditor nation. As a trading nation, it is the world's third largest by value after the United States and Germany. Its phenomenal growth rate compared with all other industrialized countries in the world has been driven by its exports of advanced technological products with an increasing reputation for quality and reliability. At the same time manufacturing industry in Japan has achieved a high level of productivity faced with competition not only from the United States and Europe but also from lower cost economies in South East Asia and the Pacific Basin. Japanese manufacturers are adapting to new circumstances by shifting into new sectors and by investing heavily in research and development for this purpose. In addition they are investing in overseas production centres to capture new markets and avoid the charge of protectionism. In stark contrast to the United Kingdom, whose manufacturing sector has been declining, Japan's manufacturing sector has been growing during the 1980s.

West Germany

As we have already seen from the tables above, West Germany is the world's third largest economy, the backbone of which is its manufacturing sector. As in other industrialized countries, there has been a decline in traditional smokestack industries in favour of new high technology ones. It is also noticeable that until recently the manufacturing sector's share of total output was declining whilst the service sector was becoming more important. The United Kingdom and the United States have been faced with competition from West Germany over the last 30 years mainly in the fields of engineering, vehicle building, electrical engineering, precision instruments and optical goods, chemicals and iron and steel[8]. Since the end of the Second World War it has been commonplace to read of the 'German economic miracle'. In the 1950s and 1960s West Germany's economic growth averaged 6.5% per year – a figure comparable to that of Japan[9].

The significance of Germany as a competitor in world markets in the future is difficult to assess because of the problems arising from unification. In 1990 the GDP of East Germany was estimated to be only about 10% of that of West Germany[9, p.2]. Unification has created severe adjustment problems for the new Germany including a loss of output and rising unemployment in the east as a consequence of restructuring the economy. In the west, there has been a rise in inflation and in interest rates as well as higher taxes, all of which could affect competitiveness and levels of growth in the future.

1.1.4 Implications

This then is the context in which we ask the question 'Can manufacturing industry survive?' We have chosen to look at the growth trends of the world's major economies, over the longer term rather than to focus on what may be temporary fluctuations of recent years. What becomes obvious is that unless something radical and fundamental happens, unless governments and managers of our manufacturing organizations are prepared to devise new competitive strategies to ensure survival in the global marketplace of the future, then the answer to our question is 'NO'. Manufacturing industry in the United Kingdom and in the United States has declined, is declining and will continue to decline if we carry on as we have done so far. The damage that has already been done to the manufacturing base of both these economies will take a long time to repair. Not only do we have to halt the decline, we have to put things into reverse and accelerate at a pace which will prevent us from being swept downstream by the raging torrent of international competitive pressures, otherwise there will be no survival. But survival requires the combined efforts of government and industry. On the one hand, industry needs the support of government to create the right conditions in which it can flourish, and the power of governments to remove obstacles to growth which are beyond the capabilities and responsibilities of individual companies. On the other hand, government needs the support of industry to ensure its growth strategies are implemented. No matter what government policies are introduced, whether they be fiscal or monetary measures to stimulate investment, reduce interest rates and inflation or reduce taxation, at the end of the day it is the managers of our industrial organizations who make the decisions and control the processes by which improvements eventually come about. Politicians and the bureaucrats have a significant part to play, but it is to the managers whose decisions are so crucial, that we address ourselves.

We have so far concentrated on manufacturing industry because this is where the crisis has arisen. However, the 1990s recession has been as unkind to services as it has to manufacturing. Although the service sector grew substantially in the 1980s, it is not without competitive pressure. For example, in 1964, the top ten commercial banks were six American, two Canadian and two British, ranked by value of deposits. By 1988, the top ten were all Japanese.

Other evidence of the penetration of traditional domestic markets may be seen in French contractors undertaking refuse collection, Japanese companies owning golf clubs and European companies taking over haulage contraction.

It cannot by any means be assumed that the service sector is safe. The approach to our problems must encompass all aspects of our economy. What managers can do to improve the competitiveness of their organizations, whether in the manufacturing or service sector, is what this book is about.

1.1.5 The fight for survival

The fight for survival has already begun in some companies. The search for better ways of doing things begins only with a realization that there is a problem to be solved. Fortunately many 'leading edge' companies have recognized the nature of the problem and have begun the agonizing, soul searching and painstaking process of defining and analysing what needs to be done to pull themselves out of the mire. These are the companies which stand a chance of survival. Others may have recognized there is a problem but have failed to take the necessary and appropriate actions to ensure survival. Too many of our manufacturing companies have 'gone to the wall' because they lacked effective management. They lacked managers capable of understanding what was required of them in terms of leadership and the ability to formulate and implement a survival strategy. We see too often, senior managers either unwilling to grasp the nettle or not knowing which nettle to grasp. How many of us have come across managers who mouth the right words but only pay lip service to new policies, new procedures, new solutions: managers who refuse to recognize that it is themselves who have to make the effort, as much, if not more so than the rest of the workforce?

Changes in the style of management, changes in the culture of our organizations must take place if the decline is to be stopped and put into reverse. Anyone should be able to see this yet the inexorable process of decline continues. Half-hearted efforts and superficial commitment to the improvement process will not do. Nothing short of revolution in the practices of management will do the trick. Managers if they don't know clearly what to do, are going to need to learn fast and then do it.

Let us, therefore, look at what is wrong with traditional ways and what needs to be done to improve them.

1.2 THE TRADITIONAL APPROACH TO QUALITY MANAGEMENT

1.2.1 Inspection and rejection

Looking back at some of the best known books on management written 30 years ago, it's interesting to note that the subject of quality received little attention. For example, in *The Practice of Management* by Peter

Drucker[10] there is no mention at all in the index of 'quality'. Some other texts do mention it but only in the context of quality control. This is defined by Fetter and quoted in Johnson, Kast and Rosenzweig[11] as 'the function of ensuring that the attributes of the product conform to prescribed standards and that their relationships are maintained'. The emphasis of the quality control process was on product inspection and rejection and one of the crucial issues of the time was determining when and what number of products to inspect. This in turn, involved a consideration of the features of the product that needed to be controlled and the costs involved. The inspection process might have been a 100% check of every item or a random sample based on statistical calculations.

The responsibility for ensuring that products conformed to the specifications belonged in the first instance to the quality department. However, the quality manager usually reported directly either to the production manager or the plant manager both of whom were, more than likely, faced with conflicting objectives. On the one hand they were under pressure to meet production targets in terms of number of products 'out of the door', whilst on the other hand they were required to deliver products which conformed to the specification. Where a choice had to be made, it was often in favour of letting the faulty goods go in order to meet the desired production target. The consequence of this, however, was an increase in customer complaints or customers who did not complain but did not come back again, together with the problems of having to rectify errors in the field and a rise in warranty costs. At the same time, managers were generally tolerant of high scrap levels and rework inefficiencies. The concept of an acceptable quality level (AQL) was commonly used. This meant that it was perfectly acceptable for a percentage of a lot, say 5%, to fall below the specified quality level. The quality inspector's job was to identify and record 'crime'. Some organizations did address the question 'To what extent should they also have a role in initiating corrective action?' but only in the most sophisticated companies was their role in 'crime' prevention even considered.

Confirmation that the emphasis of the quality function was on inspection can be obtained, curiously enough, by looking at the original name, in the United Kingdom, of the Institute of Quality Assurance. Today, not many people are aware that it began its life as The Institute of Engineering Inspection. Its change of name is an interesting reflection of the changing emphasis of quality management over time. It demonstrated the realization that the whole concept of quality based on inspection was not only inadequate, but wrong. As Sir Frederick Warner, formerly President of the Institute of Quality Assurance so succinctly put it, 'If you have to inspect a product once it has been produced then you have done it wrong'.[12]

This doesn't of course automatically follow. There may be some very good reasons why it is necessary to inspect a product, e.g. because of the high degree of risks involved in its use afterwards. Moreover, following inspection, it may be revealed that the product was not in fact 'done wrong' at all. Despite this, Sir Frederick's point brings home the need for the design of processes which produce products right first time, ideally, without the need for inspection. It was the overemphasis on the inspection process that was wrong.

Likewise, the definition of quality as 'conformance to the specification' was also wrong or inadequate. This assumed that the specification itself was correct or as it should be, and that products made to the specification, would satisfy the needs of the customer. Clearly in many cases this was a false assumption because the design of the product was frequently 'producer driven' rather than 'customer driven'. Customers were given what the designers and the engineers thought was best for them, not necessarily what the customers themselves, actually wanted.

If the name change of the Institute of Engineering Inspection symbolized a change in approach, what was the focus of the new orthodoxy of Quality Assurance at that time?

1.2.2 Quality Assurance

The term Quality Assurance means different things to different people. Gryna[13] defines it as 'the activity of providing the evidence needed to establish confidence, among all concerned, that the quality function is being effectively performed'. What kind of evidence do each of us need to give us this confidence? As consumers we might be content with examining the product ourselves simply by touching it or inspecting it. We might want to go further and test it ourselves by putting it to use or we might be prepared to accept the word of others that the product has been tried and tested and found to meet expectations. Usually we rely on the manufacturers to provide us with a product that they have satisfactorily tested and expect them to provide objective evidence that the product meets certain performance standards.

The concept of performance standards needs elaboration. Standards are central to any discussion of quality or quality assurance. When inspectors examine a product they have either subjective or objective criteria in their minds against which the product is being evaluated. These criteria represent the acceptable standard, and when deviations or variations from it occur, the inspector sounds the alarm bell. These two apparently simple notions of standards and variability have given birth to a substantial amount of literature about the processes involved in defining, measuring and achieving desired levels of quality in the goods and services that we consume. Warner[12] provides us with a fascinating insight into the historical background to quality assurance. Referring to the need for organizations to have an understanding of applied statistics and particularly the statistics of variability, he went on to say:

> I can always recall the Guru of Quality Assurance, who is the old American Deming, who was responsible for setting up the whole of the Japanese programme of quality after the war and responsible for the great advance they made to Quality Circles. Deming told me that the great influence in his life was when he came to England before the war and he went to University College London and he did statistics in the Department of Statistics there under Fisher and Pearson.
>
> Now they were two statisticians who refused to talk to each other. They occupied separate floors in the same department but refused to talk. Fisher was the pioneer in the application of statistics to complex problems, particularly in the field of agriculture where you have very variable properties in your materials and where you need to have some way in which you can bring together all the varying information which arises from your measurements. So it was Fisher who developed really the application of statistics to random operations in biology. He was trying to see that experiments were properly conducted so that you could draw valid conclusions. He dealt with the particular problems which arise with wide variations in results and in plotting, for example, how you would lay out an experiment so that statistically you wouldn't have interference by organizing the plot on a Latin square, that kind of thing.
>
> That was Fisher's contribution. Pearson's contribution was a different one, and it was related again to the statistics by which, if you try to set a given target then of course you're always in a distribution curve in which so many things will lie within 95% and others will lie outside, and his studies were devoted to seeing how you can avoid getting outside the limits of what you're aiming at for your approved variability in the process. This particular work we took advantage of in this country during the war when all the men were being called up and women had to be trained in the factories, and particularly in engineering factories. They were given control charts, and these were drawn up by reference to a British Standard called British Standard 600R which was devised by Pearson. This gave very simple control lines on which, in fact, you inspected products and the moment they deviated and you saw points coming outside, then you called the attention of your foreman, your manager, toolsetter, whoever it was, and they made the necessary correction so as to get the product back on to line.

Reference was made in the above extract to British Standard 600R. This is just one of thousands of standards which have been formalized and agreed either for organizations to conform to on a voluntary basis or in some cases on a statutory basis. In Britain the British Standards Institute plays a considerable role in the development of standards for the whole spectrum of industrial and commercial activities. Much of the impetus for the development of standards comes from the Ministry of Defence with what became known as 'Def Stan', followed by an appropriate number. For example Def Stan 9000 is a standard relevant to Quality Assurance in electronic components.

The majority of these standards relate to individual products and components of products or services. However, from 1979 a new standard came into effect which related to the Quality Management System as a whole rather than to the products of the system. The new standard known as BS5750 in the United Kingdom was revised more or less simultaneously in 1987 with ISO9000 which is the International Standard and EN29000 which is the European standard, so that they are now identical and all published together. From 1995 the title was changed to BS EN ISO9000. In future it is likely that they will all simply be referred to as ISO9000. Throughout most of this book, we will simply refer to ISO9000, but in all cases this can be taken to be synonymous with BS EN ISO9000 and EN 29000. In the United Kingdom the administration of this Quality Assurance Scheme is supervised by a National Accreditation Council for Certification Bodies. At the time of writing there are approximately 32 certification bodies including The British Standards Institute, Lloyds Register and Yarsleys who audit an individual company's Quality Management System to assess whether or not it meets the standard. The Institute of Quality Assurance organizes the arrangements for the provision of independent, qualified and registered auditors who carry out the assessments for the certification bodies.

Certification of an organization's Quality Management System enables it to publicize its achievement and this is often regarded as a marketing tool. It indicates to potential customers that the organization has taken the trouble to get its systems and procedures in order, and up to an objectively assessed standard such that its customers can have the confidence that the quality function is being effectively performed. Customers should be able to rest assured that the organization will deliver what it says it will deliver. This does not automatically guarantee that the product or service provided will be of a high standard. It does mean, however, that if the customer has specified a set of requirements then these should be met. It remains for the customers to determine whether or not the specification meets their expectations or whether the product is fit for the purpose for which they wish it to be used. Whilst ISO9000 is a universally recognized standard, and more and more companies are seeking and achieving certification, there are other standards of a similar kind which have been developed by individual organizations. Perhaps the best known of these is the AQAP (1–4) standard which companies supplying to the Ministry of Defence were required to meet. From September 1991 the MOD decided to replace this Quality Management System Standard with ISO9000. However, individual companies still operate their own standards. For example Ford Motor

Company has its Q1 and Q101 standards which suppliers are required to comply with if they wish to remain approved suppliers to that company.

So we have progressed from the old idea of inspection as the sole weapon in the quality management process and developed more comprehensive Quality Assurance processes to provide the protection against quality problems that might arise, and to provide customers with the confidence that things are being done properly.

The trouble with all of this though is that it is still not enough. The whole approach is only a first step towards the creation of organizations capable of producing quality products or services that meet customer requirements at a price they can afford. Companies which believe that the attainment of the ISO9000 standard is the end of the journey in building a quality organization are very much mistaken. The approach is still an old-fashioned one full of inherent weaknesses. Many companies with certification to the ISO9000 standard still operate the traditional quality control inspection processes, making products to meet the specifications which they themselves may have determined and have no processes for winning over the 'hearts and minds' of employees in a constant struggle to bring about continuous improvements either in the products themselves or the processes which are employed. Also in these companies, there is frequently very little focus on the needs of customers. Indeed the 20 elements of ISO9000 make very little mention of them at all.

If we are going to compete effectively in the global marketplace more needs to be done than to rest on the laurels of ISO9000 achievement. A new philosophy is required.

Listen to Bill Collard[14], formerly Director, Product Quality, Ford of Europe on the realization that Ford was beginning to lose out in world markets.

> We began at the beginning, by redefining 'Quality' as 'meeting the customer's needs and expectations at a price he is prepared to pay'.
>
> Historically Quality for Ford had been the achievement of specification. Our systems were set up to control to that, and we expected parts and assemblies built to specification to satisfy our customers. Our prime measure of customer acceptance was warranty returns and that was supported by our in-house data on scrap, rejects and service, parts, sales and by our own auditing and product testing, but always related to the achievement of specification.
>
> The weaknesses are obvious.
>
> Rejects and scrap became institutionalised. The concept of an AQL accepts and uses a proportion of defective material. Warranty is not necessarily customer sensitive, it is cost prioritised, not ordered according to customer dissatisfaction. Warranty often does not identify

concerns which really upset customers, because they cannot be fixed. And there are no competitive comparisons to be made because warranty data is kept very confidential by everyone.

As a matter of policy we directed ourselves to control processes rather than prices. We determined that our processes should be engineered to prevent the manufacture of defectives rather than accept the possibility of defectives and put in systems to hopefully catch them. And Statistical Process Control was accepted as the means for achieving it.

Ford's change of direction in managing the quality process came about as a consequence of looking closely at what the Japanese were doing. Some dramatic changes were occurring in the marketplace with respect to Japanese competition. For example in the mid 1970s Ford in the United Kingdom was exporting about a 1000 vehicle sets a day KD (i.e. knocked down), 600 of which were for the Asia/Pacific region[14]. By the end of the decade it had reduced to a trickle and now it is virtually non-existent. Local overseas companies could get a better quality, more reliable vehicle at lower cost from Japan. At the same time the Japanese were trying to penetrate unprotected European markets, especially Scandinavia, Belgium, Switzerland and Iceland. France and Italy were protected and the United Kingdom also took steps to restrict imports to avoid being swamped. In the United States the story was the same. Chrysler only avoided collapse with help from the US government. Ford itself incurred over four billion dollars trading losses over three years in adapting to the new situation[14].

What happened to the motor industry is happening in other sectors too, particularly in electronics. The motor industry, however, is still the world's largest manufacturing industry, the influence of which, over the rest of our manufacturing industry, is very considerable. If the motor industry is at risk in the survival stakes then so too are the rest of our manufacturing activities.

To face up to the stark and painful realization that Ford's survival was in question, they went, as we have seen, back to square one and began the new era by identifying what they meant by 'quality'. In this they were influenced by the views of Dr Edwards Deming, whom they hired as a consultant in 1982. Deming as we shall see in the next chapter was a pioneer in advocating a new philosophy now known as Total Quality Management.

1.3 THE NEED FOR A NEW APPROACH

The turbulence of the world in which we live, the pressures for change to which our industries are subjected; the advances in technology; demographic changes; the creation of new trading blocs such as the Single

European Market; the opening up of central and eastern Europe; deregulation and a host of other influences are cause enough for managers of our manufacturing industries to think deeply about the need for a new approach. Wasn't it 'Alice through the Looking Glass' who said 'Everything around me has changed so I must change too'? We have to run very hard up the 'down' escalator merely to stand still. Much more effort is required to get to where we want to be. We have agreed that British and American manufacturing industry is in decline and in danger of being eliminated by competition, not only from Japan but also from Taiwan, South Korea, Germany and elsewhere. If we accept the challenge and fight back to re-establish our competitive position, fundamental changes are required. No single panacea exists. Our theme is that Total Quality Management is a new approach which is essential to the survival of all organizations, not just our manufacturing companies. We have chosen to focus our attention in this chapter on manufacturing industry for several reasons. It is the basis upon which most of our wealth is created, it accounts for a significant part of our GDP, and it has been very severely hit by competition.

We do not contend that Total Quality Management is the only ingredient required in the new formula to restore our economic competitiveness. In our view it is a necessary ingredient but not a sufficient one. Many other strands need to be woven into corporate survival strategies over the next decade. Senior managers, government officials, politicians, trade unionists and academics all need to collaborate in the process of designing and implementing new and better solutions to the problems we face. We believe, however, that Total Quality Management is something that everyone with an interest in our economic survival can do something about. It's an approach which involves us all in searching for improved performance, more efficiency and greater effectiveness.

What then is Total Quality Management all about? In what way is it different from the traditional approach of Quality Assurance? What evidence is there that it works? These are some of the questions to which we must now address ourselves.

REFERENCES

[1] Keegan, W 'The Awful Inheritance of Eleven Mean Years'. *Observer* 25th November 1990

[2] Innovation in Manufacturing Industry, Report to Select Committee on Science and Technology, March 1991

[3] Economist Intelligence Unit, Country Profile U.K. 1992–1993

[4] 'Strengthening Growth and Improving the European Economy Conference' Annual Economic Report 1991–1992 No.50 Commission of the European Communities Brussels 1991

[5] Prowse, M 'Is America in Decline?' *Harvard Business Review* July/August 1992
[6] Economist Intelligence Unit, Country Profile U.S.A. 1991–1992
[7] OECD Economic Survey for USA 1990–1991, Basic Statistics, International Comparisons, August 1991
[8] Economist Intelligence Unit, Country Profile W. Germany 1990–1991
[9] Economic Briefing 'The World Economy'. H. M. Treasury No. 3, November 1991
[10] Drucker, P (1961) '*The Practice of Management*', Mercury Books, William Heinemann, London
[11] Johnson, Richard A, Kast, Fremont E and Rosenzweig, James E (1970) *Organisation and Management – A Systems Approach*, McGraw Hill, New York
[12] Warner, Sir Frederick, Keynote Speech to first BQA Regional Briefing on TQM, Danbury Park Management Centre, 29 June 1989
[13] Gryna, Frank M (1988) Quality Assurance Section 9.2 in Juran's *Quality Control Handbook* 4th edition (eds J M Juran and Frank M Gryna) McGraw Hill, New York
[14] Collard, W E. Keynote Address 'Planning for Total Quality Excellence' first TQM Conference held at Danbury Park Management Centre, October 1986

Total Quality Management 2

2.1 TQM DEFINED

Frequently in our discussions with managers we hear statements like 'Surely TQM is nothing new – it's simply a common sense way of remaining in business' or 'any well run company is already doing TQM without even realizing it'. Recently a survey amongst employees in a well known British company which has invested heavily in TQM over the past five years revealed that most of them thought TQM was simply what they had been taught by the company over the years but packaged differently. In the same company there was an inherent feeling that TQM was just a passing whim of the senior management and that it would be forgotten after a short period of time[1]. And yet despite these impressions and misunderstandings 'Total Quality Management' according to a report in The *Economist*[2] 'is still the Japanese trick western firms are keenest to copy'. Apparently there are now around three-quarters of American and British firms claiming to have some form of total quality programme.

What then do we mean by TQM? Is it something new or merely the same old ideas recycled? If it is a different approach, what are its essential features, what makes it different from the old ways of doing things? Total Quality Management has been defined and its essential features described in a variety of ways. There is no single universally acceptable definition – a blueprint in heaven – to which managers throughout the world can subscribe. Different protagonists have emphasized different aspects of the approach but no individual author can claim a monopoly of the truth or copyright of the set of concepts and ideas which contribute to the overall philosophy and practice of TQM.

Despite these differences, however, some definitions have emerged which have proved useful as a starting point for understanding what the central notions of TQM are all about. For example, in the United Kingdom the first serious attempt to provide an 'official' definition was made by the British Quality Association, but this was not until 1989. Their view was that

> Total Quality Management (TQM) is a corporate business management philosophy which recognises that customer needs and business goals are inseparable. It is applicable within both industry and commerce.
>
> It ensures maximum effectiveness and efficiency within a business and secures commercial leadership by putting in place processes and systems which will promote excellence, prevent errors and ensure that every aspect of the business is aligned to customer needs and the advancement of business goals without duplication or waste of effort.
>
> The commitment to TQM originates at the chief executive level in a business and is promoted in all human activities. The accomplishment of quality is thus achieved by personal involvement and accountability, devoted to a continuous improvement process, with measurable levels of performance by all concerned.
>
> It involves every department, function and process in a business and the active commitment of all employees to meeting customer needs. In this regard the 'customers' of each employee are separately and individually identified[3].

This may seem overly long and cumbersome but this is perhaps inevitable if all the aspects of the approach are to be encapsulated in the definition.

Some people might prefer to see a shorter version such as that of Munro-Faure and Munro-Faure 'Total Quality Management (TQM) is a proven, systematic approach to the planning and management of activities. It can be applied to any type of organisation'[4]. The trouble with such a concise definition, however, is that it does not tell us very much. For it to be meaningful, as the authors have recognized, it is necessary to outline the fundamental concepts behind TQM which we will examine later.

Another definition of TQM is included in a Department of Trade and Industry booklet entitled *Total Quality Management* prepared by Professor John Oakland. This states that 'TQM is a way of managing to improve the effectiveness, flexibility and competitiveness of a business as a whole. It applies just as much to service industries as it does to manufacturing. It involves whole companies getting organized in every department, every activity and every single person at every level'[5].

The United States Department of Defense defines TQM as 'both a philosophy and a set of guiding principles that represent the foundation of a continuously improving organization. TQM is the application of quantitative methods and human resources to improve the materials and services supplies to an organization, all the processes within an organization, and the degree to which the needs of the customer are met, now and in the future'[6].

Some further definitions which might have more appeal to practitioners because they are taken from commercial organizations, are set out below.

The first comes from the Royal Mail in which Total Quality is

> A comprehensive way of working throughout the organization which allows all employees as individuals and as teams to add value and satisfy the needs of the customer.
>
> A business-wide customer driven strategy of change which moves us progressively to an environment where a steady and continuous improvement of everything we do is a way of life.
>
> Identifying and satisfying the needs of the customer starting with the external customer and working backwards so that quality at each stage is defined in terms of the 'next customer' in the process.
>
> Being both effective (delivering the right products to the right segments of the market) and efficient (doing so at the most economical levels possible)[7].

The second is from the Total Quality Management handbook published by The British Railways Board in 1989. For them 'Total Quality Management is the process which seeks to meet and satisfy customer requirements throughout the whole chain of internal and external customers and suppliers'[8].

Cullen and Hollingum (1987) provide a very succinct statement of what Total Quality means to them. 'Total Quality means exactly what it says – zero defects in products leaving the factory and in the services offered. It means Quality in every aspect of the company's operations'[9].

All of these definitions have appeared since we ourselves attempted to define TQM back in 1988. Our view then was, and still is, that 'it is a process of individual and organizational development, the purpose of which is to increase the level of satisfaction of all those concerned with the organization: customers, suppliers, stakeholders and employees'[10].

Finally, no brief look at definitions would be complete without mention of what many people now regard as the official definition as contained in the British Standard BS5750: Part 1:1992 Section 3.1 Total Quality Management (TQM). Here it is defined as 'Management philosophy and company practices that aim to harness the human and material resources of an organization in the most effective way to achieve the objectives of the organization'. This is accompanied by the following three explanatory notes.

> NOTE 1: The objectives of an organization may include customer satisfaction, business objectives such as growth, profit or market position, or the provision of services to the community, etc. but they should always be compatible with the requirements of society whether legislated or as perceived by the organization.

NOTE 2: An organization operates within the community and may directly serve it, this may require a broad conception of the term customer.

NOTE 3: The use of this approach goes under many other names some of which are as follows:

continuous quality improvement;
total quality;
total business management;
company wide quality management;
cost effective quality management.

The Total Quality Management Standard BS7850 came into effect on 15 September 1992. It is published in two parts[11];

Part I is a Guide to Management Principles

Part II is a Guide to Quality Improvement Methods.

These two documents are exactly what they say they are – guides. We understand from the British Standards Institute that it is not their intention to certificate companies which meet the requirements of BS7850 in the way in which companies achieving BS EN ISO 9000 are certificated by nationally accredited certification bodies.

What do all these different definitions of TQM tell us? First of all, and not surprisingly, there is a common thread running through each of them, namely, the comprehensive nature of the approach. It involves all people at all levels in all functions. It really is total in every sense of the word. There is a danger, however, that in practice too much emphasis might be placed on the quality aspects of the definition or even on the management component of the definition. All three elements are of equal significance.

The second thing that these definitions reveal is that there are, within each definition, different points of emphasis. For example, the United States Department of Defense seems to emphasize the importance of the application of qualitative methods as well as human resources to the improvement process. Others, e.g. British Rail, stress the importance of satisfying customer requirements, whilst Cullen & Hollingum focus on zero defects in the product leaving the factory.

One could legitimately ask, at this point, if a product with zero defects necessarily satisfies the customer's requirements. This is a question that frequently arises in discussions about the relevance of the Quality Management System BS EN ISO 9000 to TQM and the relationship of each to satisfying customer requirements. Let us explore this in more detail.

2.1.1 The Relationship of TQM to the Quality Management Standard BS EN ISO 9000

We are often asked whether or not there is any connection between the Quality Management Standard BS EN ISO 9000 and the TQM approach. Our response to this is that it is possible for any company to be certificated to BS EN ISO 9000 without TQM, or a company can have TQM without BS EN ISO 9000. Alternatively a company can be certificated to the Standard first and then pursue a TQM approach. Some companies with which we have worked have embarked on the installation of both processes at the same time. If an organization intends to adopt both approaches, it makes sense to us for it to adopt an integrated methodology towards their implementation. The essential differences between each approach, however, need to be recognized. BS EN ISO 9000 is a Quality Management System in which the emphasis is on the writing of formal procedures and work instructions to guide employees. The expectation is that all employees will comply with the procedures in order to ensure that the work is done properly. Internal and external audits are carried out in order to identify whether or not employees are complying with the requirements, and where they are not, corrective action is taken to remedy the deficiencies. The focus, is, therefore, on the technical system and the way it operates.

But organizations are more than technical systems. Apart from anything

BS EN ISO 9000	TQM
• Not necessarily customer focused	• Definitely customer focused
• Not integrated with corporate strategy	• Integral to company stategy
• Technical systems and procedures focused	• Philosophy, concepts, tools and techniques focused
• Employee involvement not necessary	• Emphasis on employee involvement and empowerment
• No focus on continuous improvement BS EN ISO 9000 – a destination	• Continuous improvement and TQM synonymous. TQM a never ending journey
• Can be departmentally focused	• Organization wide – all departments, function and levels
• Quality Department responsible for Quality	• Everyone responsible for Quality
• More likely to preserve the status quo	• Involves process and culture change

Figure 2.1 Differences between BS EN ISO 9000 and TQM.

else, they are also social systems. They are about people and about the way people behave and interact with each other in groups. They are about the attitudes, the aspirations and the motivation of people in work situations. Quality products and services as well as quality processes can only be developed through attention to this social system. The technical system needs to be integrated with the social system in order to build a Quality Culture. Total Quality Management is about integrating these two systems through the adoption of managerial processes which provide a focus on customer needs, employee needs and the needs of the stakeholders of the organization (Figure 2.1). Companies certificated to BS EN ISO 9000 may not be so focused on identifying and satisfying customer needs, neither may they be focused so sharply on the involvement and empowerment of employees in pursuit of continuous improvement. See our 'three-legged stool' model in Chapter 4.

In a Total Quality Management approach, there is a need for managerial leadership to create the appropriate characteristics of a Total Quality Culture. Management need to ensure appropriate structures are installed to manage the improvement process and they need to provide the necessary resources for improvement to take place. The relationship between these three elements – the technical system, the social system and the managerial system, is shown in Figure 2.2.

2.2 ORIGINS AND GROWTH OF THE CONCEPTS OF TQM

At the beginning of this chapter not only did we pose the question 'What is TQM?', which we have now briefly answered, we also asked 'Is it something new or merely the same old ideas recycled?' Unfortunately

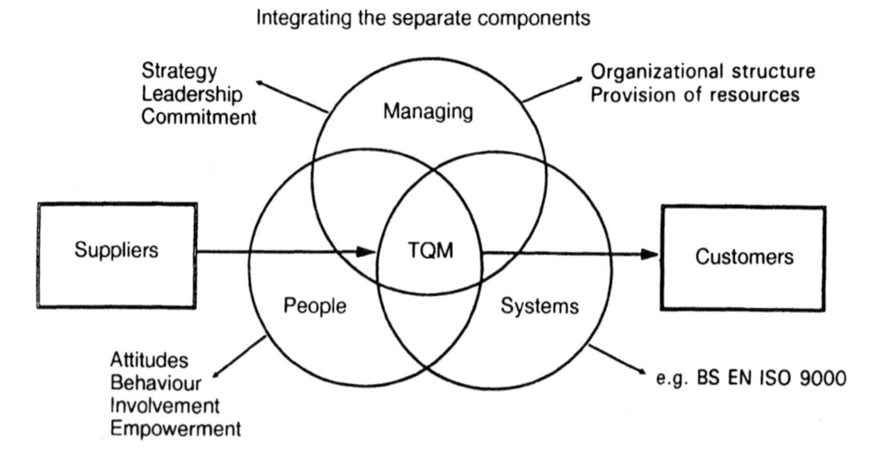

Figure 2.2 The relationship between TQM and BS EN ISO 9000

definitions alone do not allow us to answer such questions because although they may tell us what TQM is, they do not enable us to fully understand what the philosophy or methodology is behind these definitions or to see in what way, if any, it differs from previous ideas of management. To get a better understanding we need to go back to the origins and growth of the approach itself.

Many people have contributed to the development of the Total Quality Management approach but there are three names that stand out above the others as the best known original thinkers. These are W Edwards Deming,[12] Joseph Juran[13] and Philip Crosby.[14] Other major contributions have been made by Feigenbaum,[15] Conway[16] as well as by specialists like Taguchi,[17] Ishikawa[18] and Shigeo Shingo.[19]

2.2.1 W Edwards Deming

Deming, an American consultant, is generally regarded as the father of the TQM revolution. As a statistician during the 1940s he made a significant contribution to the improvement of quality in the United States through the application of statistical process control (SPC) to work situations. For example in 1940 he improved the US Population Census by introducing sample proofing of card punching instead of carrying out a 100% inspection process. This led to an increase in work flow by 600%. In 1942 he was invited to train engineers at Stanford University in statistical process control (SPC) in order to reduce scrap and rework levels in the production of war materials. But it was not until General Douglas McArthur invited Deming to Japan in 1950 as part of the post-war reconstruction initiative that the full significance of Deming's approach was realized.

We have already seen in Chapter One the phenomenal success of the Japanese economy over the 30-year period from 1960 to 1990. Japan's reputation for producing quality products is now legendary and the Japanese themselves are the first to acknowledge Deming's contribution to this success. This acknowledgement was first reflected in the initiation in Japan in 1951 of the Deming Application Prize (despite Deming's own antipathy to prizes and awards), and later in 1960 when Deming was decorated by Emperor Hirohito.

What are the essential elements of Deming's approach? When he was once asked to summarize his philosophy he replied 'If I have to reduce my message to management to just a few words I'd say it all had to do with reducing variation'.[20]

It should be noted at this point that Deming's recognition of the importance of reducing variation was not his own original idea. This had previously been recognized by Walter Shewhart who, when working in the

Bell Laboratories in 1931, became concerned with the economic control of the quality of manufactured products and with the formulation of criteria for determining when numerical data are in statistical control. He pointed out that:[21]

- All processes and systems exhibit variability
- Variability may be caused by either random or special causes
- Special causes must be eliminated before implementing process changes to improve productivity

Deming in developing these ideas emphasized that random or common causes of variation are inherent in the processes which managers themselves have designed and established. He estimates that as much as 94% of problems arise through systems deficiencies rather than through the fault of the operators of these systems or processes.

Realization of this has of course significant consequences for the way people are treated at work. Deming's view was that management too often blame employees for things that are beyond their control and that what is really required is a 'total transformation of the master style of management'. He believed in encouraging employee participation and in enabling them to contribute to continuous improvement through their understanding of the processes and how they can be improved.

Managers themselves often induce variation through their own behaviour, policies and practices. For example there will be variation in the supply of raw materials if purchasing is carried out by competitive tendering where the price tag is uppermost in selecting suppliers. Even if only one source of supply is used there are likely to be variations in the input. Where there are constant changes of suppliers or the more suppliers we have, the more variation we force into the process that uses these supplies.

Another example of the way managers can induce variation is the way in which they use fear as a motivator. This discourages people from admitting they do not understand things and encourages them to bury their mistakes. This does not encourage teamwork, neither does management's tolerance of divisiveness and competition within the organization and yet some managers not only tolerate this, they encourage and reward it.

Deming has written a great deal over the years to explain his philosophy and methods. Some of his key thoughts are embodied in the following distillations of his thinking:

- Deming's fourteen points
- The seven deadly diseases
- The sixteen obstacles
- The new climate
- A system of profound knowledge

Here, to illustrate the nature of his thinking let us explore the first two of the above, i.e. his fourteen points and his seven deadly diseases. Readers wishing to learn more can either read his book *Out of The Crisis*[12] or Frank Price's illuminating work on the Deming approach called *Right Every Time*[22], or alternatively they can contact the British Deming Association for more information (see Appendix A – useful addresses).

First of all, let us look at Deming's definition of quality. His view is that quality may be defined as a 'predictable degree of uniformity and dependability at low cost and suited to the market'. His fourteen points for managing quality are[12]:

1. Create constancy of purpose toward improvement of product and service, with the aim to become competitive and thus to stay in business and to provide jobs.
2. Adopt the new philosophy. We are in a new economic age. We no longer need to live with commonly accepted delays, mistakes, defective materials and defective workmanship.
3. Cease dependence on mass inspection to achieve quality. Require instead statistical evidence that quality is built in.
4. End the practice of awarding business on the basis of price tag alone. Instead, minimize total cost by working with a single supplier.
5. Improve constantly and forever every process for planning, production and service.
6. Institute modern methods of training and education on the job, including management.
7. Adopt and institute leadership.
8. Drive out fear, so that everyone may work effectively for the economy.
9. Break down barriers between staff areas.
10. Eliminate slogans, exhortations and targets for the workforce asking for zero defects and new levels of productivity. Such exhortations only create adversarial relationships, as the bulk of the causes of low quality and low productivity belong to the system and thus lie beyond the power of the workforce.
11. Eliminate numerical quotas for the workforce and numerical goals for management.
12. Remove barriers that rob people of pride of workmanship. Eliminate the annual rating or merit system.
13. Institute a vigorous programme of education and self-improvement for everyone.
14. Put everybody in the company to work to accomplish the transformation.

The deadly diseases

In his pursuit of continuous improvement and his view that the western style of management needs to be completely transformed he identified what he believes to be the key weaknesses which he calls the deadly diseases[12]. These included:

- A lack of constancy of purpose.
  To achieve continuous improvement, customer values should dominate all decision-making and day to day decisions must be consistent with the long-term plan. Confusion amongst the workforce, customers and suppliers will result if there are constant changes in direction or lack of constancy of purpose.
- Emphasis on short term goals.
  Managers lack a vision of what their organizations will look like in 5 to 10 years time and what is in it for each of the stakeholders. Too much emphasis is given to short-term profile and not enough on the needs of customers, employees, suppliers and the public.
- Evaluation of performance – merit ratings and annual reviews.
  Deming argues that performance evaluation disregards the fact that employees work with systems that are variable and unstable and hence the evaluation system is biased and inconsistent. Performance evaluation also disregards the fact that employees work in groups and the system thus mitigates against teamwork. Annual merit rating and ranking of people urgently need to be removed. Understanding causes of variation has implications for our approach to reward systems, labour relations and personnel policies.
- Mobility of management.
  Job hopping amongst managers has a detrimental effect and encourages short-term thinking. This is especially relevant at CEO/director level.
- Management only by the use of visible figures with no consideration for unknown figures.

Deming believes that the most important figures that management needs are unknown or untraceable. Examples of this that he gives include:

> The multiplying effect on sales that comes from a happy customer, and the opposite effect that comes from an unhappy customer
>
> or
>
> Loss from the annual rating of performance

These things cannot be quantified but are nevertheless significant. Neglect of the more important unknowable or invisible figures will almost certainly lead to a deterioration in the profitability of the company. In fact on visible figures, a company may appear to be doing well, but in reality could be going downhill for failing to take account of these unquantifiable factors.

The above ideas are only a brief introduction to the depth and breadth of Deming's thinking and serve to illustrate that for him quality is as much to do with people as it is with products. It is about the importance of management's role and commitment and its obsession with quality. He stresses the importance of educating employees to understand the processes employed in organizations and the causes of variation as well as the need for teamwork. Traditional approaches to management have created barriers which need to be broken down. Above all he recognized the importance of the business as a system with its interdependent parts linked in a process chain which included suppliers and all internal company functions focusing on meeting the needs of the external customers.

2.2.2 Joseph Juran

Like Deming, Joseph Juran made a significant contribution to the quality revolution in the post Second World War reconstruction of Japan. Also like Deming he was decorated by the Emperor of Japan in recognition of his contribution. Both Deming and Juran had previously been employed at the Hawthorne plant of Western Electric Company in Chicago where, between 1924 and 1932, the famous Motivation Studies carried out by Elton Mayo took place leading to the emergence of the 'Human Relations' school of management thinking and later to the Organization Development approaches in which TQM has its roots.

Again, like Deming, Juran argued that it was system deficiencies that gave rise to errors and waste rather than the mistakes made by the workforce. Juran, however, shifted his emphasis to the management of quality. He realized that quality control techniques were reasonably well developed but little emphasis had been given to how quality could be or should be managed. By 1951 Juran had formulated a fairly coherent set of ideas about the management of quality which he set out in *The Quality Control Handbook*[13]. This is now in its fourth edition and is probably the most comprehensive book on all aspects of quality: Quality Control, Quality Assurance and Quality Management, ever written.

Juran noted that the word 'quality' has multiple meanings, two of which dominate the use of the word [13, section 2.2].

> 'Quality consists of those product features which meet the needs of customers and thereby provide product satisfaction'

> 'Quality consists of freedom from deficiencies'

Acknowledging the need for a universally acceptable comprehensive definition of quality he has chosen what he believes to be the most convenient short phrase – 'fitness for use' but recognizes that this has not achieved universal acceptance. He disagrees with the view that quality can be defined as consisting of conformance to some standard – for example conformance to specifications, procedures or requirements because they do not define adequately the quality responsibilities of the company. 'For the Company the definition should be stated in terms of (1) Meeting the customer's needs and (2) freedom from deficiencies' [13, sections 2.7, 2.8].

The Juran trilogy of management processes

At the heart of Juran's thinking about managing quality is the need to present his ideas to senior managers in an easily understood form. To achieve this he has conceptualized his thoughts in a trilogy of management processes: Quality Planning, Quality Control and Quality Improvement [13, section 6.4], all of which are interrelated.

Quality Planning consists of a series of steps:

- Determine who are the customers
- Determine the needs of the customers
- Develop product features which respond to customer needs
- Develop processes which are able to produce these features
- Transfer the resulting plans to the operating forces

Quality Control
The second managerial process in Juran's trilogy is Quality Control. This is an essential process for assisting the operating forces to achieve product and process goals. Deficiencies are identified, causes analysed and problems are prevented from getting worse. Juran stresses, however, that such deficiencies arise and chronic waste occurs because the operating processes are planned that way. The quality control process consists of the following three steps:

- Evaluate actual operating performance
- Compare actual performances to goals
- Act on the difference

These are steps which the operational groups can employ with the aid of statistical control processes in order to control quality.

Waste can only be reduced or eliminated by attention to Quality Improvement, the third process in the trilogy. Essentially the Quality Planning process needs to be improved by ensuring that the quality planners have adequate time and resources to do their effective planning in the first place, but more attention is also required to be given to the feedback mechanisms available to ensure that lessons are learned from the problem-solving activities of the control process as well as during the improvement process.

Quality Improvement

The Quality Improvement process is perhaps the most significant of Juran's contributions to the TQM movement. Quality Control processes are more concerned with maintenance of a level of quality either through the prevention of errors or the correction of errors when they occur. The improvement process is at the heart of TQM. The search for never-ending or continuous improvement is what it is all about, not just in the quality of the product or service provided but also in the processes employed. Juran emphasized that the improvement of both products/services and processes applied to all customers and he was one of the first, if not the first, to recognize that customers were both internal to the organization as well as external. He defined each of these customers in the following way:

External customers. 'These are impacted by the product but are not members of the company (or other institution) which produces the product. External customers include clients who buy the product, government regulating bodies, the public (which may be impacted because of unsafe products or damage to the environment) etc.'

Internal customers. 'Within any company there are numerous situations in which departments and persons supply products to each other. The recipients are often called "customers" despite the fact that they are not customers in the dictionary sense – i.e. they are not clients'.[13]

The notion of the customer-supply chain has played an important part in the development of quality improvement approaches, especially since meeting customer needs is at the core of the whole philosophy. The exercise of trying to define who one's customer is, in itself can be a very practical and useful one. It is not always as easy and as straightforward as it seems. But if there is no clear perception of who the customers are in the first place, then it is more likely that there will not be a clear perception of their needs either. Time invested in trying to sort out workflows and processes, key input variables and key output variables, is essential if the situation in which the 'tail is wagging the dog' is to be avoided. The concept of the internal and external customer supply chain shifts the emphasis away from the traditional or scientific model of hierarchical and functionally oriented perspectives towards a more dynamic process oriented model.

In order to set about improving quality, Juran has formulated a series of ten steps which companies can follow. These are dealt with in the next chapter on approaches to implementation of TQM.

Above we have briefly summarized some of Juran's key contributions.

Limitations of space prevent us from doing full justice to both Deming and Juran and so we need to move on to our third major contributor, Philip Crosby.

2.2.3 Philip Crosby

In many discussions about Total Quality Management one of the concepts frequently referred to is that of 'zero defects'. This is a Crosby concept introduced by him in the early 1960s when he was employed by the Martin Corporation, in charge of the Pershing Missile project. Crosby is founder of Philip Crosby Associates which he set up in 1979 and which is now owned by Proudfoot Inc. He is author of many books including *Quality is Free*,[23] *Quality Without Tears*,[24] *Running Things*[25] and *The Eternally Successful Organization*,[26] in which he explains more fully his philosophy and approach. Many of his ideas about quality improvement are derived from his 14 years of experience from 1965–1979 when he was Corporate Vice President with responsibility for quality at ITT.

The notion of 'zero defects' is a controversial one but forms an integral part of Crosby's belief in the four 'absolutes of quality' [26, p.35].

The four absolutes are a simple but effective means of addressing the TQM philosophy and can be applied to any situation in which one finds oneself in business or organizational life. The first absolute is that everyone must understand quality in the same terms, that is as 'conformance to the agreed requirements of the customer', rather than as goodness or excellence. The 'customer' here is expanded to take in the concept of the 'internal customer', i.e. anyone to whom work is passed in a chain of internal processes which ultimately lead to the external customer. The important point is that quality must be ensured at each stage if the end customer is to receive a satisfactory product or service.

Secondly, there must be a system to ensure quality (conformance). That system is concerned with preventing errors, not checking or appraising them. Thirdly, the standard of performance against which non-conformance must be measured is zero defects, not acceptable quality levels or percentage defective values. While the level of non-conformance is greater than zero there must still be room for improvement.

Crosby's fourth absolute answers the question of how quality should be measured. He calls this the price of non-conformance, preferring the word 'price' to 'cost' since 'you have a choice as to how much you pay', depending on your organization's attention to the other three absolutes. Crosby asserts that 'manufacturing companies spend at least 25% of sales doing things wrong; service companies spend at least 40% of their operating costs on the same wasteful actions'[26].

As we have seen, Juran disagrees with the definition of quality as 'conformance to requirements' since he regards it as inadequate. As for the concept of 'zero defects' many people find this unacceptable either because it is seen as an unrealistic target to achieve and hence can become dysfunctional, or because it is seen as an exhortation to the workforce – a means of motivating them to avoid making mistakes.

Crosby maintains that his concept was misrepresented. It is intended to be a performance standard to be applied to all processes. He sees it as part of an attitude of mind. When things go wrong we should not regard them as inevitable, rather we should aim to prevent them from occurring in future so that we eventually achieve defect-free production of goods or services. The concept of zero defects contrasts with the notion of Acceptable Quality Levels which suggests that we can set a target of say 95% or 98% defect-free supplies and accept the remaining faults. Zero defects implies that it is worth investing more time and money to prevent the remaining faults. As Konasuke Matsushita, President of the Matsushita Corporation says 'it is better to aim at perfection and miss than to aim at imperfection and hit'[27].

Neither Deming nor Juran accept the standard of zero defects. Juran rejects it more so than Deming, since he believes that there is a law of diminishing returns on quality and that a point can be reached where further improvements in quality are more expensive than tolerating a level of failure. Deming is particularly scathing about the use of zero defects as a slogan targeted at the workforce, pointing out that the individual worker only has limited control over the factors which can cause quality failings and that management must take the primary burden of building in quality to the systems, tooling and materials the worker has to use.

Crosby attributes the majority of quality problems to management in much the same way as both Deming and Juran do. His estimate is that 80% of the problems are caused by management and thus the cure for these problems lies with management leadership. In fact the role of management is crucial. In his book *The Eternally Successful Organization* Crosby stresses the fact that whilst it is necessary for a company to conduct formal education and training in order to build an implementation process for Quality Improvement, 'the essential ingredient is executive integrity. When the big boss just will not tolerate shoddy performance and is a personal example of what should occur then the company has a chance'[26, p. 33]. He believes that 'companies reflect the standards of their leaders and the tools of quality will not alter those standards'. What is required for permanent improvement is [26, pp. 35–36]:

- The conviction by senior managers that they have had enough of quality being a problem and want to turn it into an asset.
- The commitment that they understand and implement the Four Absolutes of Quality Management. They have to accept the responsibility for making this happen. The quality department cannot do it.

- The conversion to that way of thinking on a permanent basis. This replaces the conventional wisdom that caused the problem in the first place.

Some of Crosby's other views worth noting if change in organizations is to occur are[26]:

- People will take quality just as seriously as management takes it, no more.
- Integrity is unrelenting; it can't be done in short bursts of enthusiasm stemming from regret.
- The tools of quality control are not designed to cause prevention throughout the organization.
- Think about Quality Improvement in terms of earnings per share. A well established process will double it.
- Every individual in the company needs continual education concerning his or her role in getting things done right the first time and clear requirements on the changing scene within which we live and operate.

2.3 TQM AND TRADITIONAL MANAGEMENT THINKING

Many of today's managerial practices, the way organizations are structured, their policies, their procedures, their styles of managerial behaviour and other manifestations of organizational cultures are a legacy of the management theories of an age gone by.

The evolution of our thinking about organization and management is depicted in Figure 2.3. What this shows is the way in which our ideas of management grew originally in the main out of the practice of the Church and the military. This was followed by the contributions of Weber, Fayol and Taylor which gave rise to Scientific Management and the Classical School. A turning point came in 1929 with the work of Elton Mayo at the Hawthorne plant of the Western Electric Company when the Human Relations School was born. This was the beginnings of the application of behavioural science to the study of organizations. Meanwhile Scientific Management continued to develop on a parallel path in what might be regarded as the application of Management Sciences to the study of organizations and management.

With Bertalanffy's[28] work we see the emergence of Systems Thinking which was an integrative force leading us into contingency theories of management in the 1970s and then into TQM which has dominated the scene during the 1980s and now into the 1990s. We saw earlier how both Deming and Juran were employed in the Hawthorne Plant of Western Electric Limited at the time when the Human Relations School and later the Organization Development approaches arising from behavioural

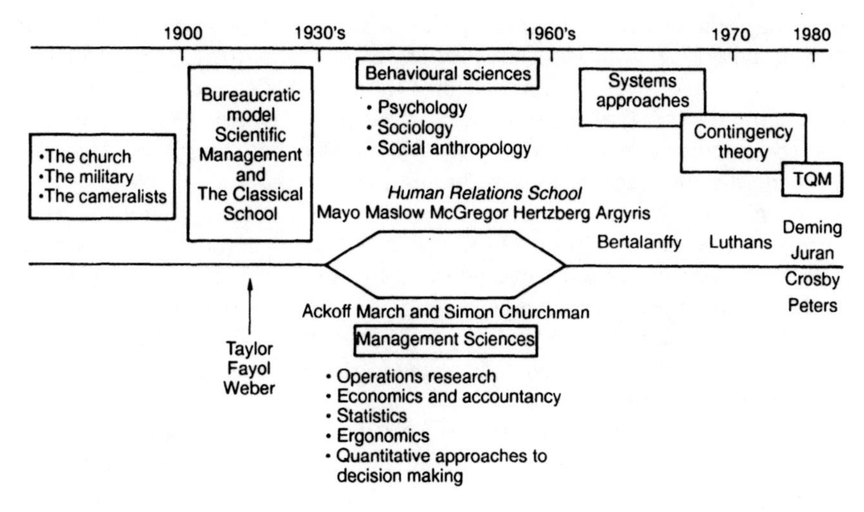

Figure 2.3 TQM and the development of management thought.

science applications began to emerge. In our view the seeds of the TMQ movement were sown all these years ago. It has taken more than 50 years for the plant to grow and take root in the west.

We still suffer from simplistic notions about how to organize and manage people at work based on narrow or even false assumptions of human motivation prevalent in the 1920s. The consequences of rigid adherence to the precepts of the 'Scientific Management School' have rendered many organizations inflexible, unresponsive to change and riddled with dysfunctional approaches to improving efficiency and effectiveness.

As we said in Chapter One, our economic performance in world markets has been abysmal. Companies have gone bankrupt, and the percentage of imports has risen both in the United States and in the United Kingdom. The need for radical transformation has never been more apparent than now, if we in the west wish to survive the ever-increasing competition from the Pacific Basin countries.

Total Quality Management has, as we have seen, grown out of the adoption of Deming and Juran's ideas in Japan since the war. Deming and Juran themselves have been influenced by the contributions to management theory of the behavioural and management sciences, particularly systems thinking, which have steadily been taking hold since the end of the Second World War. But the rate of progress in the west in recognizing the need for new thinking and implementing new practices has generally been very slow. It was almost 30 years after the Japanese began to introduce the principles of Total Quality Management that leading companies in the USA began to adopt similar methods. As far back as 1967 Juran predicted that 'the Japanese are headed for world Quality Leadership within the next two decades'[29]. And if the United States was slow to respond, Europe has been even slower. It was not until the mid-1980s that any significant developments in TQM started to occur. Now in the early 1990s there are few companies unaware of the TQM philosophy, but the success rate for implementation of the ideas is disappointing.

The *Economist* report referred to previously, quotes the findings of a survey carried out by Arthur D Little of 500 American manufacturing and service companies. This discovered that only a third of them felt that their TQM programmes were having a 'significant impact' on their competitiveness[2]. At the same time a study by A T Kearney of over 100 British firms revealed that only 20% surveyed, believed that their quality programmes had achieved 'tangible results'[30].

The *Economist* report also points out that 'of those programmes that have been in place in western firms for more than two years, two thirds simply ground to a halt because of their failure to produce hoped for results'[2]. The reasons for this are not entirely clear but there are some indicators. John Cottrell, Manager of A T Kearney, notes that of those companies surveyed, those which succeeded with their quality programmes shared four common characteristics[30]:

- An emphasis on tangible results
- An insistence on performance measurement
- An integrated programme
- A clear commitment from top management

In other words, unless organizations are willing to face up to the radical transformation of their existing practices, unless they are willing to do more than play with the idea, pay lip service to it and only make superficial attempts to change things, they are unlikely to get better and their half-hearted efforts will be a waste of time.

We are talking about fundamental changes in the culture of our organizations, not just quick fix solutions that can be grafted on overnight. Such transformation takes time, often more time than many managers first anticipate. The slowness of change may result in disillusionment, despondency and a desire to give up. Despite this, where the desire to change exists, the questions managers need to ask are: Where do we begin this process of long-term change? How do we make the breakthrough required to set out on the path of successful organization development?

2.4 TQM AS AN ORGANIZATION DEVELOPMENT PROCESS

Organizations change whether they like it or not. Some changes occur in response to external pressures. For example, when the Ministry of Defence decreed that all its suppliers must be certificated to the BS5750/IS09000

Quality Management Standard, managers reacted to this requirement to make the necessary changes if they wished to remain suppliers to the MOD. Likewise managers react to legislation to ensure that they are not in breach of the law.

Not all change, however, is reactive. Change can occur naturally – that is as a result of a series of incremental decisions taken by different managers in different parts of the organization with no grandiose plan or scheme of things to guide them. Change can be haphazard or it can be planned. Many of the changes which have occurred in organizations have had no coherent strategy underpinning them and often changes in one part of the organization are inconsistent with changes in other parts of the organization. Often the fragmented structures and the uncoordinated decision-making processes that prevail make this inevitable.

Some changes that occur in organizations are neither reactive nor natural, they are the result of the proactive stances of managers deliberately planning and implementing changes in accord with a preconceived strategy to bring about improvements. This is the nature of a properly managed TQM installation process.

Many authors have written about managing change,[31] others have focused on Organization Development[32] and a third group have been concerned with TQM. Very few have seen TQM as a highly focused form of Organization Development but in our view this is precisely what it is. It is the process of changing an organization's culture or developing the organization in such a way as to make it more responsive to customer needs on the one hand, whilst becoming so in a more efficient and effective way on the other. To become more responsive to customer needs, changes in the way organizations are managed are likely to be essential.

The traditional cultural characteristics of organizations derived from Taylor's model and other 'Scientific Management' principles need to be changed in a planned and systematic way. At the same time managerial constraints and behavioural patterns also need to be changed. For those who are not familiar with the fundamentals of Taylor's model or with Scientific Management in general, a brief outline might be helpful.

Frederick Winslow Taylor (1856–1915)[33] spent his whole life dedicated to the study of work in search of the most efficient way in which individual work tasks could be performed. Using time and motion methods, he gave birth to the notion of 'objective' standards of performance alongside which an incentive scheme was devised for rewarding employees who performed above standard and penalized employees who produced sub-standard work.

Managers took responsibility for planning, organizing and supervising work and determined the best way for jobs to be done – 'each man receives in most cases complete written instructions describing in detail the task which he is to accomplish. The task specifies not only what is to be done but how it should be done and the exact time allowed for doing it'. In return the man was paid for his efforts.

Such an approach failed to recognize the more complex variables affecting human motivation, and yet this job engineering approach has dominated manufacturing industry for most of this century. Only in recent years has there been any significant questioning of the approach and the emergence of alternatives to job design such as job enrichment, quality circles and empowerment of the workforce.

Resulting from these kinds of traditional beliefs about managing organizations, many have developed as 'punishment-centred bureaucracies' where employees are assumed not to want to give of their best or get it right first time and, therefore, need to be coerced or controlled as well as exhorted to do a good job. In such circumstances where employees are not trusted, managers feel inclined to increase the controls over staff.

The result may lead to a vicious circle in which employees react against the controls and become demotivated, whilst management decide to increase the controls even more. Managers then begin to ask the wrong question – 'How do we motivate our employees?' instead of 'How do we avoid demotivating them?' In a Total Quality Culture one of the keys to success has been the empowerment of employees, giving them the power, not only to be responsible for quality work, but to take appropriate actions when things go wrong or to take preventive action before they go wrong. A cultural shift away from firefighting to a prevention orientation is thus required in which everyone can take responsibility for making sure errors are avoided. It is not just management's job as argued by Taylor.

Other cultural characteristics include the need for open communications between employees at all levels, instead of having a communication pattern based on the grapevine and secrecy. Many managers regard the possession of information as a source of power and in a desire to preserve their power base, hold on to that information. This mitigates against the creation of a quality culture in which the needs of the customers are paramount rather than the needs of managers to empire build or put their own personal needs first.

Frequently, because managers are often more concerned with their own needs for stability and security or to protect their own vested interests, there is a resistance to the continuous change and improvement demanded by a Total Quality Culture.

Overcoming resistance to change, especially when it requires a change in the behaviour of managers is one of the greatest reported difficulties faced by organizations attempting to undertake TQM. As we shall see later, for any implementation process of TQM to be successful, an appropriate strategy for managing the change process will need to be adopted. This strategy should recognize the likelihood of resistance to such change and have built into it the means for overcoming such resistance. Whilst successful implementation requires top management committment, it cannot simply be imposed from above without the commitment from people at all levels throughout the organization.

2.5 INGREDIENTS OF SUCCESS

Earlier we reported some of the disappointments experienced by companies attempting to introduce TQM into their organizations and suggested that reasons for failure were not altogether clear. One question which might legitimately be asked is 'is TQM itself, as a philosophy or methodology, a failure, or is it to do with the way in which the processes involved are implemented?' We believe it has a lot to do with the latter. In 1989 the Develin Report into 'The Effectiveness of Quality Improvement Programmes in British Business'[34] observed that the most successful companies have four outstanding elements:

- Careful planning and execution of the process
- Middle management involvement from the outset
- Deliberate targeting of quick tangible results (fast track)
- Constant employee communication, involvement and recognition

The more successful companies also rated all aspects of TQM as important but emphasize:

- Clear mission and constancy of purpose
- Clear policies towards customers, suppliers and employees
- Clear critical success factors
- Right culture and attitude towards quality
- Clear responsibilities with ownership of processes

2.6 BENEFITS FROM SUCCESSFUL TQM INSTALLATIONS

Despite the negative comments contained in the *Economist* report to which we have previously alluded, many companies involved in the pursuit of excellence through Total Quality processes have reported quite staggering results. Perhaps one of the most successful stories comes from Rank Xerox.

Rank Xerox were once synonymous with copiers. During the 1970s, fierce competition from Japan drastically reduced their market; they slipped from over 18% of the US market share in 1979 to 10% in 1984. By 1980 the unit manufacturing costs of Rank Xerox were more than double those of their Japanese competitors. Coupled with slower delivery and poorer quality, Rank Xerox were forced to act to survive. Their initial response was traditional – across the board cutbacks and closures. This simply reduced quality further and despite pricing improvements they could not win back their lost markets. The company turned to a TQM strategy.

They went back to basics and adopted a prevention-based approach to their business. They introduced supplier assessments, benchmarking – a system of setting targets for each process in the business to improve to the best in that process, regardless of industry sector, massive training and employee involvement/worker control, and 'just-in-time' manufacture. The direct to indirect staff ratio reduced from 1.4:1 in 1980 to 0.4:1 in 1988 and the number of inspectors was reduced to one-eighth of the number they had in 1980. This was achieved without compulsory redundancy and many executives were helped to set up their own business serving Rank Xerox as suppliers. They introduced more flexible work methods including home-based operations and moved their reward systems to directly reflect customer satisfaction. Senior executives became responsible for processes rather than functions and are now directly rewarded by reference to the ratings customers give them on their performance.

Successes reported include:

- Overhead costs on products reduced from 360% to 137%
- Customer satisfaction improved by 38%
- Service response cut by 2 hours (27%)
- Manufacturing costs halved
- Product quality up by 93% from 91 defects per 100 machines to less than 7
- 40% reduction in unplanned maintenance
- Inventory reduced from 90 to 14 days of supply
- 40% improvement in delivery times
- In book-keeping, errors in supplier payments reduced from 1 per 100 to 1 per 1000 – the benchmark is 1 per 1 000 000
- 80% inspection on incoming parts reduced to 15% (primarily new parts or suppliers) – reject rates on supplied parts have fallen from 30 000 per million to 300 per million (the benchmark is 100 per million)

Unlike many other household names that have entered terminal decline, Rank Xerox's commitment to TQM has turned the company around and enabled it to regain its place among the world leaders. By 1988 it had recovered to 13.8% of the US market and increased profit from £166 million before tax to £415 million. In 1990, Xerox copiers topped 5 of 6 categories in customer satisfaction expressed through industry surveys[35].

Other success stories abound. We have collected at least 50 examples illustrating how companies have benefited in significant ways from their TQM initiatives. Unfortunately we cannot list them all, but a selection of these success stories are outlined below which serve to show the range of benefits.

2.6.1 Reduction in customer complaints

Avery's European Materials group gave complaint management top priority. If a customer is not happy Avery provide a free replacement or credit note. At the same time they use diagnostic and statistical tools to improve manufacturing and design processes which cause the complaints in the first place. In 1984 the company paid out 1.4% of sales as the cost of complaints; this reduced to 0.9% by 1988 – a saving of 1.9 million dollars. In 1983, customer complaints took more than 7 days to settle in 75% of cases. By 1988 only 1 in 5 were over 7 days in processing. Paperwork is reduced as a by-product and cash flow improves through less delays in payment. Avery initially invested 1 million dollars in training and implementing TQM; in 5 years the number of complaints has reduced by 45% in a business which has doubled in size, returns and allowances have reduced by 40%, 70% less credit notes are issued whilst scrap has gone down by 35%. Absenteeism is down by 25% and the cost of quality has been reduced by 25% – representing 15 million dollars in operating costs[36].

2.6.2 Reduction in costs of quality

Caterpillar's quality process is set to reduce their costs of non-conformance from 400 million dollars to less than 80 million over 7 years. Each of Caterpillar's units has a Quality Committee which operates under a six-point approach:

- assign responsibility for the project
- define the customer/supplier relationships and tasks
- determine quality measures and control points
- evaluate process information
- simplify processes as far as possible
- apply SPC and prevention systems to control the process

Caterpillar estimate that for every £1 spent on quality measures, £10 is saved[37].

2.6.3 Increased market share and reduced costs

Domino, a Cambridge-based company specializing in electronic overprinting, went into TQM to support their efforts in achieving BS5750. In Japan they now have 25% of the market share, having taken customers from such giants as Hitachi. Their initial estimated cost of quality was £2.19 million of their profit. This has been substantially reduced, on-time delivery is up from 75% to 95%, and supplier returns down to 0.5%. Of all the small businesses which originally came to the Cambridge Science Park, Domino is the only one to make it to the Stock Market without being rescued or taken over. They are currently either first or second in all of their markets[38].

2.6.4 Reduced complaints and accidents

Florida Power and Light were the first non-Japanese company to win the Deming Award in 1989. Their Quality Improvement Process has, amongst other things, helped to reduce the average length of consumer service interruptions from 100 minutes in 1982 to 48 minutes in 1988, halved the number of complaints and reduced accidents to employees by over one-third. The price of their electricity has been held below the cost of living and is now less than it was in 1985[39].

2.6.5 Reduced defects – increased customer satisfaction

Ford USA, who got into TQM in the US in 1979, have reduced their number of parts shipment locations from 2750 in 1984 to 2000 in 1989 with a target of around 1500 in 1992. Since 1980, half of its retained suppliers have been approved for their quality systems, covering 60% of Ford's total supplies. By 1992, 95% of all supplies should be from approved suppliers under the Q1 rating system. On warranty repairs per 100 vehicles, the number of incidents has been reduced from over 500 things wrong per 100 vehicles in 1980 to 180 things wrong in 1988. Over the same period their customer satisfaction rating has risen from 66% to 86%. Ford's goal for 1992 is to get down to 80 things wrong per 100 vehicles and 95% customer satisfaction. If achieved this would put Ford as world leader on quality in vehicle manufacture (assuming the rest don't improve further than they expect)[40].

2.6.6 Increased efficiency

Girobank moved into TQM in 1988. In the first year 28 registered improvement projects were started. By 1990, 1500 had been registered and over 300 completed, yielding total savings of £4m. In 3.5 years the company has seen the following benefits:

- 40% reduction in inventory
- 80% reduction in errors in operations directorate
- 94% of customer transactions processed on day of receipt, up from 57%
- 66% fewer customer complaints over errors
- 12% fewer operations staff
- BS5750 accreditation from March 1990

Through a supplier quality improvement partnership with Post Office Counters, errors at Post Offices have been reduced by 75%[41].

2.6.7 Increased profit, productivity and market share

Hewlett Packard report increased profit by 244%, productivity by 120% and market share by 193%. At the same time they reduced stock by 36%, manufacturing costs by 42% and failure rate by 79%[42].

2.6.8 Multiple benefits

IBM's corporate headquarters at Rochester, USA, is the hub of its Total Quality Process. Among many exceptional aspects of its performance and commitment to quality are[43]:

- 30% improvement in productivity since 1986
- manufacturing cycle time reduced by 60%
- three fold increase in product reliability
- capital spending required for defect detection reduced by 75%
- write-offs as a proportion of output down by 55%
- 5% of payroll spent on training
- $3.6 million paid out in quality awards to 40% of its staff in 1989
- over 350 teams working on benchmarking its quality standards
- over 1000 supplier employees trained in Quality Improvement techniques such as SPC, design of experiments and JIT by its own staff in 5 years
- six senior managers having ownership of improvement in one of its six critical success factors which are:
  - improved product and service requirements definition
  - enhanced product strategy
  - six sigma reliability
  - further cycle time reductions
  - improved education
  - employee involvement and ownership

2.6.9 Increased sales, reduced costs, reduction in cycle times

In 1986 when ICL commenced its TQM process, 65% of its customers said they would recommend them to others. By 1987 that figure had risen to 89%, more than any of its main competitors. It has achieved a 30% reduction in manufacturing costs and a reduction of £100 million in inventory on sales revenue of £1.3 billion. Inventory turn has risen from 3 in 1985 to 7 in 1988. Cost of quality has been reduced by 10% at Kidsgrove, saving them £9 million in 2 years. Design to manufacture cycle time has been reduced from 76 weeks to 26 weeks and the manufacturing cycle itself from 5 weeks to just 3 days[44].

2.6.10 Savings from improved communications and co-operation

Pirelli reckon to have saved almost £11 million by the improved communication and co-operation which TQM brings. Labour turnover has dropped from 20% to 0.5% and plant utilization has increased from 35% to 75%. External waste has dropped by 73% and stock turn has improved from 4.4 to 13.7[45].

2.6.11 Start up and production costs reduced

Toyota implemented over 7 years an integrated product design, engineering and manufacturing quality process called 'QFD' (Quality Function Deployment) which reduced their overall initial production costs on new products by 61%. Costs associated with start up, which had previously been over 75% of the total costs (i.e. recalls, warranty, customer complaints, design changes, etc.) were virtually eliminated. Costs were moved up front to the more intensive planning process using QFD[46].

2.6.12 Increased return on investment

Transamerica Insurance Group calculated savings of 20 million dollars in under 2 years of two special TQM-based improvement activities. They reckon on 7 to 1 return on their investment in the process[47].

2.7 SOME RECENT REPORTS ON MEASURING SUCCESS

All of the above examples are drawn from individual companies reporting the benefits which they have achieved. Now let us turn to recent reports which have surveyed, in each case, a variety of companies. We will look specifically at the range of measures of success which these surveys have identified. The Kearney report referred to earlier stated that only 20% of the respondents reported tangible results. What is perhaps more significant, however, is that 80% of the respondents reported that they did not actually measure their improvements. Consequently they were unable to state whether or not they had achieved tangible results. This is really very surprising.

Common sense tells us that the only way to assess whether or not the actions which are taken to bring about improvements are working is to measure the results. For example, if people wish to reduce their weight they embark on a programme of exercise or diet or both. The assessment of the effectiveness of the programme is measured on the weighing scales and possibly with a tape measure as well.

Why is it then that companies are reluctant to measure the effectiveness of their Total Quality interventions to see whether they have embarked on the most appropriate courses of action? Is it because they don't know what to measure, or because they find it too difficult? This is a question to which we do not know the answer. We do know, however, that those who do make the effort, use a range of measures. For example, the Kearney study[48] identified the following financial and non-financial measures employed by companies in its survey:

Financial measures

- Annual sales/revenue/income
- Stock turnover ratio
- New business won
- Return on capital employed
- Net profit
- Sales per employee
- Added value per employee
- Product costs
- Warranty costs

Non-financial measures

- Average absenteeism rate
- Lead times
- Annual labour turnover rate
- New product introductions (no. per year)
- On time delivery
- % defective products

Another study undertaken by Professor John Oakland[49] and his colleagues at the University of Bradford set out to establish whether or not similar patterns of behaviour were emerging in European companies as was shown to be emerging in American companies through the US General Accounts Office Survey with regard to establishing links between TQM practices and bottom line results. The study placed emphasis on the positive drive for quality improvements in satisfying the end customer. The business performance indicators or measures were chosen to reflect business performance in both the short and the long term and were financial measures only.

Financial measures

- Profit per employee
- Profit as a % of sales
- Profit as a % of assets
- Sales per employee
- Fixed assets - current year as a % of fixed assets - previous five years
- Pay per employee
- Number of employees trend

The Bradford Study does not suggest that TQM leads directly to improvements in bottom line results. However, the consistency of the results does point to a strong association between them. A high proportion of the companies surveyed exhibited above industry average performance against all the financial indicators measured.

The study also found that the number of employees trend is not conclusive but 17 of the 29 companies studied have, over the five-year period of the review, increased rather than shed staff.

In *Making Quality Work: Lessons from Europe's Leading Companies*, Binney [50] adopted a case study approach to analyse in depth the real-life experiences of Total Quality implementation. His aim was to dispel the myths and confusions about Total Quality in order to get it back on track. The misinterpretation of the findings of the Kearney Report did a lot to undermine people's confidence in the TQ philosophy and approach. Since there are many success stories a more balanced analysis was required. The study itself adopted the performance measures employed by the US General Accounts Office as listed below:

Financial measures

- Sales per employee
- Return on assets
- Return on sales
- Cost of quality
- Cost savings

Non-financial measures

Employee related:

- Attendance
- Labour turnover
- Safety
- Health
- Suggestions received

Operations related:

- Reliability
- On-time delivery
- Order processing time
- Errors/defects
- Product lead times
- Inventory turnover

Customer related:

- Customer satisfaction
- Complaints
- Retention
- Market share

The survey concluded that the focus of Total Quality has matured over time. It is now more concerned with broad organizational and cultural change to achieve continuous improvement in performance. It further concluded that bringing together real customer focus fact-based management and the unlocking of people potential is becoming a condition of survival.

If it is true that there has been an increasing emphasis on organizational and cultural change it is not surprising that Personnel or Human Resource specialists have become more deeply interested and involved in the implementation of Total Quality processes. One recent study, *Quality: people management matters* [51], sets out to identify the differences in experience between Human Resource professionals operating in different sectors and industries as well as the differences in involvement of Human Resource people in different types of quality initiatives. This study asked participants to rank in order the criteria used to measure the success of the programme. The means identified were:

Financial measures

- Profitability
- Unit costs

Non-financial measures

- Market share
- Quality of service/product
- External customer satisfaction
- Internal customer satisfaction
- Employee satisfaction
- Employee participation

The study concluded that there is a role for the Human Resource function in the implementation of Total Quality other than their operational role – that is, changing attitudes, generating commitment, etc. Recognition of the Human Resource function in this role is increasing but the role has previously been small as programmes have been seen as process driven. The report also noted that different sectors treat the involvement of the Human Resource function differently and this may affect their selection of criteria for success.

2.7.1 Conclusion

What the above review of some recent survey reports has shown, along with the review of the reported benefits from individual companies, is that TQM unquestionably works. The individual companies have produced hard evidence of the benefits which they have achieved from their TQM implementations. The surveys have also shown a range of financial and non-financial measures which can be employed to provide evidence of success. What is also very clear is that without resort to measures of any kind, it is impossible to say whether the implementation process is producing benefits. People may believe their actions are worthwhile but this is nothing more than an act of faith. What gets measured gets done. Measurement is thus a cause of success as well as a barometer of that success.

Our argument is, therefore, that where companies have experienced failure, or where the successes have not been as expected, rather than discuss the philosophy of TQM, look to the way in which it has been implemented. Make sure that the essential ingredients of success are present. We firmly believe that the key to success is mainly to do with the way the implementation process is handled.

In the next chapter we will examine a variety of approaches to implementation, some of which are likely to lead to failure and some of which are more likely to lead to success.

REFERENCES

[1] Reproduced with kind permission from a confidential Student Assignment, Diploma in Management Studies, Anglia Business School, Danbury, UK, 1993

[2] 'The Cracks in Quality', *The Economist*, April 18th 1992, pp. 85–86

[3] British Quality Association Newsletter (1989), British Quality Association, London

[4] Munro-Faure, Lesley and Munroe-Faure, Malcolm (1992) *Implementing Total Quality Management*, Financial Times/Pitman Publishing, London

[5] Oakland, John (1991) Total Quality Management, Department of Trade & Industry, London

[6] Total Quality Management Master Plan, Department of Defense USA, Washington, August 1988, p. 1

[7] Royal Mail internal document, reproduced with kind permission of Keith Harrison, Quality Director, Royal Mail Eastern Region, Chelmsford 1993

[8] David, M (1992) 'Is the Quality Management System as Implemented by the telecommunications Department of British Rail a Positive Contribution to TQM?' MBA Dissertation, Anglia Business School, Danbury, UK

[9] Cullen, Joe and Hollingum, J (1987) *Implementing Total Quality*, IFS Publications Ltd, Kempston, Bedford, UK

[10] Pike, R J and Barnes, R J (1988) The Total Quality Management Research and Development Centre, Anglia Business School

[11] BS7850 Part 1 1992, Total Quality Management Section 3.1 p. 3, BSI Standards, London

[12] Deming, W. Edwards (1982) *Out of the Crisis*, Cambridge University Press, Cambridge, UK

[13] Juran, J M and Gryna, F M (1988) *Quality Control Handbook*, 4th edition, McGraw Hill, New York

[14] Crosby, Philip B (1979) *Quality is Free*, McGraw Hill, New York

[15] Feigenbaum, Armand V (1986) Total Quality Control, 3rd edition, McGraw Hill, New York

[16] Conway, William E, President and Chief Executive Officer, Nashua Corporation

[17] Taguchi, Genichi (1987) System of Experimental Design, New York Unipub, Kraus International Publications

[18] Ishikawa, Kaoru (1985) *What is Total Quality Control? The Japanese Way*, Prentice Hall, Englewood Cliffs, NJ.

[19] Shingo Shigeo (1992) *Zero Quality Control: Source Inspection and the Poka-Yoke System*. Translated by Productivity Inc., available from Productivity Europe Limited, 6 West Street, Olney MK46 5HR, UK

[20] Deming, W Edwards, Reproduced from Conference Papers, Leatherhead Food Research Association Conference, June 1992

[21] Shewhart, W (1931) *Economic Control of Quality of Manufactured Products*, Van Nostrand, New York

[22] Price, Frank (1990) *Right Every Time Using the Deming Approach*, Gower Publishing Company, Aldershot, UK

[23] Crosby, Philip B (1979) *Quality is Free*, McGraw Hill, New York
[24] Crosby, Philip B (1984) *Quality Without Tears*, McGraw Hill, New York
[25] Crosby, Philip B (1986) *Running Things*, McGraw Hill, New York
[26] Crosby, Philip B (1988) *The Eternally Successful Organization*, McGraw Hill, New York
[27] Matsushita, K
[28] Bertalanffy, Ludwig Von, 'General Systems Theory: A New Approach to the Unity of Science', *Human Biology*, December 1951, pp. 303–361.
[29] Quoted in Conference Papers 'A Commitment to Operational Excellence and Continuous Improvement' Exxon Co International, Houston, Texas, May 7–9 1990
[30] Cottrell, John. 'Favourable Recipe', *The TQM Magazine*, IFS Publications, Kempston, February 1992, p. 17
[31] Bennis, Warren G, Benne, Kenneth D and Chin, Robert (1969), *The Planning of Change*, 2nd edition, Holt, Rinehart and Winston, London
Connor, E Patrick and Lake, Linda K (1988), *Managing Organizational Change*, Praegar, New York
[32] Beckhard, Richard (1969), *Organization Development, Strategies & Models*, Addison Wesley, Reading, Mass
Bennis, Warren G. (1969), *Organization Development, Its Nature, Origins & Prospects*, Addison Wesley, Reading, Mass
Woodcock, Mike and Francis, Dave (1981) *Organization Development Through Teambuilding*, Gower., Aldershot, UK
[33] Taylor, Frederick Winslow (1911), *The Principles of Scientific Management*, New York, Harper & Brothers
[34] Develin & Partners, 'The Effectiveness of Quality Improvement Programs in British Business', London 1989
[35] Rank Xerox, various sources
[36] *TQM Magazine*, IFS Publications, Kempston, August 1989
[37] *TQM Magazine*, November 1988
[38] *TQM Magazine*, August 1989 and Daily Mail 22nd September 1991
[39] *TQM Magazine*, February/August 1990
[40] Video, Litton Industries
[41] *TQM Magazine*, August 1991
[42] *TQM Magazine*, November 1988
[43] *TQM Magazine*, February 1991
[44] *TQM Magazine*, November 1988 and Develin report, September 1989, and DTI Report 1900
[45] Develin report, September 1989
[46] *Quality Progress*, American Society for Quality Control, Milwaukee, June 1986
[47] Quality and Productivity Association Conference Papers – Sept 1990
[48] A T Kearney (1991) *Total Quality: Time to Take Off The Rose-Tinted Spectacles*, London (in association with *TQM Magazine*).
[49] Oakland, J *et al*. (1993) 'The Bradford Study', University of Bradford Management Centre.
[50] Binney, George (1992) *Making Quality Work: Lessons From Europe's Leading Companies*, Economist Intelligence Unit, London, October.
[51] Baron, A and Marchington, M (in association with Wilkinson, A and Dale, B) (1993) *Quality: People Management Matters*, Short Runs Press for the Institute of Personnel Management and UMIST, Exeter.

Approaches to the implementation of TQM

3

3.1 THE NEED FOR REVOLUTIONARY CHANGE

Tom Peters' book *Thriving on Chaos* is subtitled *Handbook for a Management Revolution*[1]. Peters acknowledges the fact that the word 'revolution' is unpalatable and troubles businessmen, but insists that nothing short of a revolution will enable the western world to overcome the threat to its competitiveness position.

The idea of a revolution suggests the overthrow of the status quo; of the establishment; of traditional ideas and practices. It represents a fundamental challenge to the basic assumptions upon which our traditional organizational cultures and methods have been built. Revolution conjures up the notion of fear. Fear as heads roll; fear as the old world crumbles; fear as people become displaced and disoriented; fear arising from the threat of conflict, of struggles between opposing forces and hard fought win–lose battles. No wonder businessmen don't like the thought of it! But can we escape the need? Does all change have to be revolutionary? Does all revolutionary change have to be unpalatable? What are the alternatives?

Many of those who advocate the need for change usually think in terms of evolutionary change – a slow, steady, incremental and painless process, almost imperceptible to those involved. But this is precisely what has been happening already and clearly it isn't sufficient. We have failed to retain our position in world markets, and in the league table of leading world economies. If the decline is to be brought to a halt, much less put into reverse, then the pace of change needs to be increased. Such a pace of change required can only be called revolutionary.

Despite some widespread antipathy to the notion of revolutionary change, we have seen periods of revolutionary change in the past. We only have to think back to the eighteenth century when our society was rapidly transformed from a rural, agriculturally based economy to that of an urban, industrially based one. We refer to this period as the 'Industrial

Revolution' – a period of social dislocation in which the patterns of peoples' lives were noticeably transformed within a short space of time, but also a period of enormous wealth creation upon which our political democracies and standards of living were founded.

What managers, with the support of politicians, now need to do is to create another revolution, a revolution in which there is a rapid transformation of organizational processes based on the philosophy of Total Quality Management, Continuous Performance Improvement, or whatever other label people wish to use.

In creating this revolution, we can learn some lessons from the past. The trick will be to foster a revolution in which everyone wishes to participate, one which avoids the perceived or actual negative consequences of previous revolutions and one which promotes the benefits for all. Moreover we need to ensure that the revolution will lead to better ways of doing things rather than simply replacing old ways of doing things with new ways of doing things, which are no better, or even worse. In other words we must avoid change for the sake of change. We must also ensure that we avoid using the old ways of doing things to try to introduce the changes that are required. The characteristics of the new culture need to infuse the processes that are employed to bring about that new culture. This point needs to be stressed. To create a Total Quality Culture within an organization needs an implementation process which itself exhibits TQM principles and precepts. Or to put it another way, managers who do not fully understand the nature of TQM and what a Total Quality Organization would be like, are unlikely to be able to create the appropriate conditions and adopt the most appropriate strategies required for the implementation process to be effective. Half-baked implementation processes will lead to half-baked results. Could this then be the reason why so many attempts to install Total Quality have been unsuccessful? We believe it is.

Too many attempts have been characterized by over simplistic models of the change process. Social, political, technological, economic and cultural change within an organization is a lengthy, complex and sometimes tedious process in which the rate of advance in one part might vary from the rate of advance in another part. If this happens it will make it even more difficult to create a company wide Total Quality Culture and differential approaches tailored to overcome the barriers to change will need to be devised.

What kinds of change strategies do companies employ? From our observations and research, we have identified a number of approaches to implementation. These are:

- The managerial decree approach
- The managerial sales campaign approach
- The spread of knowledge or education only approach

- The Quality Circles or problem-solving only approach
- Standard implementation or 'packaged' approaches
- The tailored implementation or planned change approach.

3.2 THE MANAGERIAL DECREE APPROACH

It has often been stated that implementation has to start at the top of the organization. The justification for this proposition is that, presumably, unless the managing director or chief executive takes the initiative no-one else will or that if someone else does, without the managing director's support, it will be doomed to failure.

This is, of course, a nonsense. Not all change has to come from the top downwards. In fact a great deal of change may come from pressure from below upwards, or from the middle outwards. In an authoritarian culture it may well be the case that managers only act when the top person tells them to do so. Managers may always be looking over their shoulders for approval or some sign from the boss that a particular course of action is acceptable. But a managerial decree, a letter or a memo from a senior manager to junior managers or other levels of employees, announcing a new policy can equally be doomed to failure if the policy is unacceptable to those concerned. There is also a risk that if the senior manager leaves or is promoted elsewhere, the whole process may fall flat. It is true that the top manager can act as one of the biggest barriers to change and, therefore, it is desirable that his or her commitment to the policy is obtained, but so too can middle managers act as barriers to change. Their commitment is also needed to change.

If the top managers or management team have no means of enforcing the decree, it is unlikely that the policy will be implemented unless others believe in it themselves. There is also, inherent in this situation, the paradox of a senior manager unilaterally announcing that the company is going to adopt a TQM strategy, the essence of which is about involvement or participation of others in decision-making processes.

Managerial leadership is a vital ingredient in getting the process started and in keeping it going afterwards, but the decree from above by itself is unlikely to be an adequate approach for a successful implementation of TQM in the long run and thus will not reap the full benefits. Potential benefits may occur, but without the full commitment of managers and employees at all levels, the initiative will need to be sustained with further pronouncements, threats or exhortations to keep it going.

The experience of Gates Hydraulics Limited may be interpreted as an example of the managerial decree approach. In April 1988 the managing director of the company issued the policy pronouncement [2, pp. 129–30] shown in Figure 3.1.

TOTAL QUALITY MANAGEMENT

'QUALITY' is the key operating priority in Gates Hydraulics. Our goal is to give Quality Top Priority, Top Status and Top Dedication in every action and decision we take. We all think that we understand the subject and are convinced that our ways are right. However, few of us would like to explain it and discussions on the subject of Quality are usually short and superficial.

WHY CHANGE?

The pursuit of Quality must be recognized as the most important objective within the company. Quality is a prime determinant of competitiveness in contract awards, retention and overall profitability. I have, therefore, decided to adopt a policy of 'Total Quality' within our company under my direction.

The factors affecting product quality are so numerous that it is difficult to identify an area of management that is not involved in some way. Quality influences are to be found not only in the technical areas of design and production, but also in marketing, purchasing, personnel, finance and indeed every sector of the company's activities.

Quality planning must not be considered in isolation, but must be approached in the context of overall management planning, which is concerned with the corporate long-range plan for profitable growth.

It follows that all this demands clear quality objectives being set, of which the prime objective must be to achieve a high degree of customer satisfaction with due regard to quality costs. Here 'quality' must include any aspect of the product or service for the customer.

'Get it right first time' must be the golden rule for each and every department of the company if the aim is to operate cost-effective Quality Management. Basic changes of attitude will be required affecting all departments and levels within the company, including myself and fellow Directors. This may be like putting new wine into old caskets and is likely to necessitate an extensive review and possibly revision of our existing management systems.

This announcement is to inform you of my immediate objectives. You will shortly receive more information about the plans and objectives relative to individual departments from your functional Directors and Managers.

Through Total Quality Management, our goal will be a target of 90 per cent reduction in defects in products and service in three years, with a 25 per cent reduction in the first 18 months.'

A.J. Roberts
26 April 1988

Figure 3.1 Example of managerial decree – memo.

This decree reveals an interpretation of TQM as an improvement process in response to the need to change in order to remain competitive. What it also reveals, however, is an interpretation that the way to implement this improvement process is through a traditional management by objectives (MBO) approach. It has all the hallmarks of a top-down authoritarian style of management which the TQM approach suggests is contrary to effective long-term performance improvement. There is thus an inherent contradiction in this method of implementation.

That this approach did not work as effectively as it was hoped, is evidenced by the memo, shown in Figure 3.2, issued by the quality director 18 months later [2, pp. 140–41].

This second memo is more in the form of an attempt to sell or persuade people to act. This takes us into the next method of implementation.

3.3 THE MANAGERIAL SALES CAMPAIGN APPROACH

This approach is not too far removed from the managerial decree approach. Here the power to decide organizational policy still seems to rest only with senior management. They may have come together to consider the advantages and disadvantages of embarking on a TQM approach, and may have reached the conclusion that it is an appropriate direction in which to go. If they assume that it is their job to make the decisions, the next question they might ask is 'How are we going to sell this to the rest of the workforce?' The answer frequently appears to be 'Let's have a campaign'. It is not unnatural for managers to adopt this approach if they work in an organization in which the tradition has been to make things and then attempt to sell them to customers. The underlying assumption of this approach is that customer behaviour is influenced by a sales or marketing campaign. People have to be persuaded to act in the way we want them to act – by buying our products or services or by complying with our wishes as managers to carry out a certain range of tasks or activities. Selling managerial decisions to the workforce becomes part of the everyday routine of managers – manipulating, cajoling, coercing, using incentives or threats of punishment to encourage or motivate people to comply.

The cultural characteristics of organizations that emanate from this set of assumptions about human behaviour are in fact those very characteristics which the TQM approach sets out to change. The campaign approach is, therefore, unlikely to produce sustained results. The means of achieving the ends need to be consistent with the ends themselves. The campaign approach usually features memos from the senior management to the next level down reminding them of their obligations to do all they can to persuade their subordinates to 'put quality first' or 'get it right first time' (as if people set out to get it wrong first time deliberately!!). Meetings with

GATES HYDRAULICS LIMITED

MEMO

To: See Distribution List Date: 23 October 1989
From: R.A. Scouse
Copies: A.J. Roberts
I. Gardiner

TQM - Training for Quality

TQM has been recognized since May 1988, as the way forward to improving quality of product and service, together with improved efficiency. It is fair to say improvements have happened in many areas, but we are only on the fringes of the many opportunities open to the company.

We have not reviewed our progress with current objectives, neither have we prepared TQM Objectives for 1990. In November 1988 I suggested we should form a Quality Council, which would be the steering group for the TQM programme and should comprise members of the Operations Committee who should address themselves only to TQM issues. If this were successful it should reduce the need for other committees and meetings.

I do not believe we have real conviction, commitment or conversion.

- **Conviction** comes when the Management Team recognize that 'quality' is really important and decide to bring about quality improvement.
- **Commitment** comes when members of the Management Team decide they want to take actions necessary to bring about quality improvement.
- **Conversion** comes when members of the Management Team are willing to make the pursuit of quality a routine part of the company's operation.

The three C's must be accepted and I suggest we should step back and consider them so far as our operation is concerned. Once they are accepted we should embark upon a planned programme of training and work experience.

Effective training is particularly important for any TQM programme. The key element is that we cannot produce quality products or services unless we have well-trained and competent staff. Without quality products we can fall into a descending spiral, where we compete only on price, losing inevitably to our competitors. More serious, it creates a poor image, making consumers reluctant to buy and workers reluctant to join.

Nowhere in our business should we tolerate poor quality; and we cannot have quality without trained people to produce and maintain it.

This is where the Operations Committee should come in as a 'ginger' group, looking at the overall scene from a position of experience, identifying any blank areas, 'gingering' all concerned to make good the deficiency and monitoring the results.

Trying to promote plain good sense is inevitably an uphill struggle. It should not deter any of us. It should be part of our training - of ourselves and our colleagues.

Figure 3.2 Example of the sales campaign approach – memo.

exhortations to employees to take a pride in their work, or to avoid mistakes are then supplemented with poster campaigns to reinforce the message. To really drive the message home, the company training officer or quality manager may come up with the bright idea of showing groups of employees on a regular basis a series of videos about quality. Needless to say these have to be lighthearted and short because 'we cannot tax people's brains too much, they won't understand!', or so it is argued. This is another fallacy which the TQM approach sets out to remove.

3.4 THE SPREAD OF KNOWLEDGE OR EDUCATION ONLY APPROACH

Some attempts to introduce TQM into organizations are based on the view that all that is necessary to cause people to change their behaviour, is for them to be informed of what is required of them and then they will perform accordingly. This approach assumes that people are rational and that they pursue their own rational self-interest. If the benefits of adopting a particular form of behaviour are explained people will then recognize the value of the approach and consequently adopt the proposed changes. Or so the argument goes.

This strategy is a very popular one. Our universities and other educational institutions are preoccupied with research and the development of knowledge on the one hand and the diffusion of knowledge on the other. There is no doubt that many situations have been changed as a result of the spread of knowledge. But for this means of change to be effective, there is a necessary precondition. People have to be ready to receive the knowledge and ready to accept the changes which it suggests. Not all change is welcome since for some it may represent a threat to their personal positions of power, status or security or it may represent a loss in some other way.

The spread of new knowledge may only serve to create resistance to change rather than a positive, enthusiastic response and a willingness to introduce improvements. Education and the spread of knowledge may take several forms. It might explain *what* needs to be done or it might explain *how* objectives might be achieved, or it could do both. Informing people of new requirements does not guarantee that changes in practices will follow. If people do not know how to do what they have been asked to do, nothing is likely to happen.

Even where staff know what to do and how to do it as a result of some form of educational or training programme, this again may not be a sufficient set of circumstances for change to occur. There are many factors which have to be taken into consideration when explaining human behaviour. The way individual employees behave or perform at work can be represented by the equation shown in Figure 3.3.

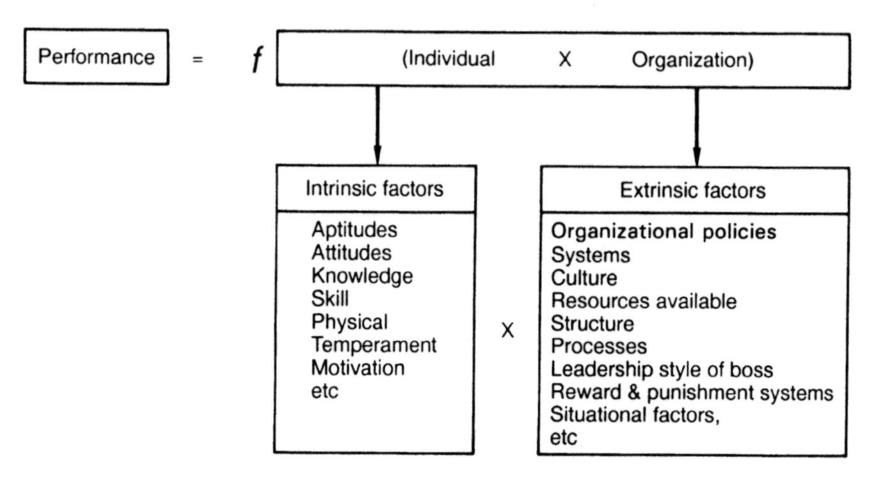

Figure 3.3 Performance as a function of a combination of variables.

What this means is that performance is a function of a combination of variables related to the individual on the one hand and the characteristics of the organization in which the individual works, on the other. So far as the individual is concerned, factors such as intelligence, aptitudes, knowledge, skills or competencies, physical characteristics, attitudes and motivation each contribute to his or her behaviour. But equally the characteristics of the organization – factors such as the organization's policies, procedures, practices, systems resource availability, leadership style, structures, communication patterns, reward and punishment systems, opportunities and many others, also play a significant part in influencing individual behaviour. We all know of examples in which an individual knows what to do, and how to do it, but is prevented by the system from doing it.

For effective change to occur, changes in the individual's knowledge or skills may need to be supported by corresponding changes in the cultural characteristics of the organization. Education alone may not be enough. Managerial action may be required to create conditions in which the newly acquired knowledge or skills can be applied.

3.5 THE QUALITY CIRCLES OR PROBLEM-SOLVING ONLY APPROACH

Quality Circles have their origins in Japan. The first Quality Circle was registered by the Union of Japanese Scientists and Engineers (JUSE) in 1962. The idea grew from a suggestion by Dr K Ishikawa in the JUSE journal *Quality Control for the Foreman* who pioneered book reading circles which were essentially study groups. These then developed a problem-solving focus in order to enable members to put theory into practice[3].

3.5.1 What is a Quality Circle?

In the Department of Trade and Industry booklet entitled *Quality Circles* the following definition is provided:

> It is a group of four to twelve people, usually coming from the same work area, who voluntarily meet on a regular basis to identify, investigate, analyse and solve their own work related problems. The Circle presents solutions to management and is often involved in implementing and maintaining them[4].

It is difficult to know how many Quality Circles are now in existence but there is no doubt that they have become very popular and have spread to many countries throughout the world. One estimate is that there are about 1 200 000 circles throughout the world with approximately 12 million members involved in them. The expectation is that this number will continue to grow over the next 10–20 years[3].

Accompanying the spread of Quality Circles has been the formation of national bodies to encourage their development. Within the United Kingdom there is a National Society for Quality Circles (NSQC) which was established in 1982, whilst similar organizations have sprung up elsewhere in the world. For example, France has its French Association of Quality Circles (AFCERQ) formed in 1981; Sweden set up the Swedish Association for Quality Circles (AGCF) in 1984. There is the Quality Circles Forum of India formed in 1982, and in South America, Brazil has an Association for the Development of Quality Circles.

On an international basis there is an Association for Quality and Productivity which was originally set up as the International Association for Quality Circles in 1978[5]. Clearly what this continuing spread of Quality Circles and of associations to promote and support them demonstrates, is that the approach has an appeal to companies as a means of solving quality problems, based on evidence that if established properly, they produce beneficial results. That is not to say that all initiatives to operate Quality Circles have been successful. Only if careful attention is paid to the way in which they are set up, and the context in which they operate is appropriate, are they likely to succeed.

Where success has been evident, a number of requirements for their effective operation have been identified.

3.5.2 Requirements for success of Quality Circles

From the literature available about successful Quality Circles we can extract the following guidelines:

- Spend time in the earlier stages making sure that everyone involved is properly informed and committed before starting the programme. The organization has to be ready to accept the full implications of the approach and recognize the organizational development aspects of the implementation process. For example management will need to be willing to accept that the approach involves trusting employees, encouraging them to take responsibility for, and become involved in, improving work processes. It requires teamwork and improved communications between management and employees, possible changes in leadership style and other cultural changes.

- Design the programme to fit the organization in terms of its structure and practices.
- Make sure there is an overall management committee or steering group chaired by a senior line manager with links to all concerned.
- Appoint at least one trained facilitator to coach and support the Quality Circle leaders.
- Make sure the circle leaders are properly trained.
- Ensure top management support is given and seen to be given.
- Make it clear that participation in a Quality Circle is voluntary.
- Ensure that operating managers are committed to the approach and make time available for circles to meet on a regular basis.
- Establish 'family' groups – i.e. invite people from the same work area or group who have a common interest in and knowledge of the work processes.
- Develop a problem solving/action oriented approach – it's not a talk shop!
- Ensure all suggestions for improvement are reviewed by the steering group and where suggestions have been approved, make sure these are implemented.
- Make sure recognition is given to circle members.
- If a company suggestion scheme has previously been in operation, make sure the two approaches are properly integrated.

3.5.3 Managing the introduction of Quality Circles

As we have previously said, before introducing Quality Circles, the readiness of the organization for a change in the way it involves its employees in the improvement process, needs to be determined. This has been likened to making sure the soil and climatic conditions are appropriate before planting seeds. The organizational climate has to be right, the circumstances facing the organization have to be supportive, employee attitudes and managerial willingness to make resources available are all factors which have to be taken into consideration.

Part of the preparatory phase might require the unfreezing of old attitudes. Managers will need to appreciate the need for change and understand the objectives of the intention to introduce Quality Circles. Organizational practices which have previously supported old attitudes and styles of managing will need to be identified and removed. The communication of information to support the new policies will be necessary.

If there has been a tradition of conflict between management and employees or where employee morale is not good, steps will need to be taken to repair this situation before embarking on a Quality Circle programme. This in itself may take a great deal of time, effort, patience and a change in attitudes and behaviour to get ready for the quality journey that lies ahead. Quality Circles are not the means by which poor employee relations can be overcome. However, they can enhance already good relationships.

Once the soil has been tilled, and the unfreezing process is showing signs of readiness for change, consideration will need to be given to how the organization is going to move forward with its Quality Circle programme. Awareness of what the programme is about by people at all levels will need to be created, and an action plan formulated to train managers, establish and train steering group members and to select and train facilitators, team leaders and team members in problem-solving processes.

Timing is all important here. Managers and other employees involved will need to be free of other pressures in order to devote the necessary amount of time to plan and introduce the programme. Introducing such an approach in the middle of a reorganization or a redundancy programme, or where the work load is particularly heavy, is not the most propitious time to do so.

Following the training of managers to ensure they understand what it is all about, and know what is required of them, they will need to consider very carefully how they intend to exercise their role in supporting the process. The Quality Circle approach is essentially a bottom up one and middle managers can be powerful agents of change or they can be significant blockers of change. Middle managers are in a position to influence supervisors and employees on the one hand and senior managers on the other – as well as their other middle management colleagues. Their role in providing support is, therefore, crucial. It might be useful for groups of middle managers to consider what they can do in practical terms to demonstrate their commitment and support. Brainstorming ways of doing this might point the way forward.

3.5.4 Some benefits of the Quality Circle approach

There are two kinds of benefits arising from employing a Quality Circle approach. There are those which come from the process itself, and those that are attributable to the results of the process. Benefits from the process may be intangible or difficult to measure but are nevertheless worthwhile.

Since the process is one of involving employees in the problem-solving process, not only does it harness the knowledge and skills of employees close to the problem situation, it also helps them to derive a sense of achievement, a feeling of self esteem, a feeling of being respected and valued, a sense of belonging more to the organization and it generates a greater sense of motivation amongst employees as a consequence. (See Chapter 6 for details of results from a nationwide survey by the Industrial Society.)

Having opportunities to contribute to the improvement of the work situation are more likely to enlist the commitment of employees to the solutions when implemented, than if these solutions are imposed from elsewhere. Quality Circles generate a feeling of ownership and generally speaking 'people don't drill holes in their own boats'. The problem-solving process is in itself an opportunity for employees to develop their skills. It presents a challenge and makes work more interesting as well as encouraging employees to take a pride in their work and be responsible for their own actions. Confirmation of these kinds of benefits can be found in company attitude surveys conducted amongst Quality Circle members.

One well known study of a company's experience of Quality Circles is that by Frank J Barra[3] who was actively involved as a facilitator in the Westinghouse programme which started as far back as 1978 and grew very rapidly from 50 pilot circles in the first year to 2000 in the next 3 years. These circles now operate in more than 200 locations of the company, ranging from small offices with less than 10 employees to large plants with thousands of employees. They operate in sales, marketing, engineering, purchasing, accounting and manufacturing, not only in the USA but also in Europe, South America and London. Referring to the results of the Westinghouse programme, Barra maintains that 'The impact of Quality Circles on the attitudes and behaviours of many mid and higher level managers is truly remarkable. Entirely new methods of communication, working relationships and management styles are emerging that no amount of conventional training could have produced',[3 p. 157]. However, not only have the attitudes and behaviour of middle managers changed, there has also been an improvement in workers' morale and communications which has pleased both managers and supervisors. The company's standard attitude survey has revealed favourable responses. Illustrative of this are these examples:[3]

- Has the Quality Circles programme made your job more enjoyable? Yes 82%
- Have you spent some of your own time (lunchtime, breaks, at home, etc.) on Quality Circle matters? Yes 88%
- Should the Quality Circles programme be continued and extended to other groups? Yes 100%

The second kind of benefits to which we referred are those which are attributable to the results of the process rather than the process itself. These results are frequently more quantifiable – either in terms of problems solved or objectives achieved, or in terms of the financial benefits – savings, cost reductions, reductions in scrap levels, improved methods of working and so on. Some examples of these are described in the Department of Trade and Industry's booklet on Quality Circles – a brief summary of these is given below:

> BSDR, the largest supplier of seat belts in the UK, based in Carlisle, report savings in one year of £100 000 from 13 Quality Circles – 'more than double the cost of everything involved in the Circle programme'.
>
> At Philips Radar Communication Systems Limited, one of the circles redesigned a carrier – a component in a printed circuit board assembly process – which not only led to a more reliable process with less rework, but also a cost saving of £5000 per annum.
>
> Rylands-Whitecross Limited report that they have 24 circles in operation. Many of these have completed several projects. Savings made include £16 000 annually from energy conservation. Another project on water saved £20 000 per year.
>
> Tioxide UK Limited introduced its Quality Circle programme in 1983 and now claim that benefits amount to over £200 000 per year as a result of a rise in chemical efficiency of the production process[4].
>
> At Westinghouse, it is interesting to note that at the time of his writing, Barra maintained that 100% of the recommendations made by Quality Circles had been approved and implemented by the company. Examples of benefits include the following:

- A reduction in the amount of overshipped material supplied to vendors giving an annual cost saving of $600 000.
- A reduction in fuel consumption for vehicles by 15% giving a projected cost saving of $100 000 over a 5-year period.
- The reduction of an operation cycle time by one week, of a deshrinking process – producing a saving of $180 000 each time the procedure is used [3, p.158].

Mike Robson, author of the book *Quality Circles* maintains that experience from all over the world indicates that the costs involved in running circles will be repaid between five and eight times on an annual basis and some companies claim payback ratios of up to 15:1 per annum[6].

These benefits clearly testify to the efficacy of the Quality Circle or problem solving approach to TQM. Other non-financial benefits are also reported which encourage company management to continue to support the establishment of circles. Even where benefits are not measured, common sense often suggests the results have been worthwhile. But despite these recorded success stories, there have been some criticisms of the approach and some disagreements as to the way to go about employing Quality Circles.

3.5.5 What is wrong with the Quality Circles or the problem-solving alone approach?

The most common reason for the ineffective employment of Quality Circles as a strategy for implementing TQM is that they do not provide the most appropriate organizational machinery for tackling the key strategic issues facing the organization. The legacy of the earlier Scientific or Classical School of management thinking is functional specialization, leading to fragmentation in decision making and interdepartmental conflict. Since one of the aims of TQM is to promote integration and co-operation between different functions of the organization, cross-functional teams are required. The most serious problems facing organizations today require multi-functional approaches. No single department or section of the organization is likely to be able to remove the causes of the problems if they lie beyond the boundaries of their own control. Since Quality Circles are departmentally or work-group focused, they are thus limited to concentrating on less strategic problems. Quality Circles, therefore, tend to work within existing structures and may reinforce these rather than challenge them. Cross-functional teams not only involve people from different parts of the organization with a contribution to make, or who have a vested interest in the outcomes of the problem-solving process, they also encourage new perspectives on organizational arrangements.

However, even cross-functional teams cannot change overall policy and systems controlled by top management, such as procurement, employment conditions, capital programmes, etc. Many of these systems and policies lie at the root of quality problems and thus require the active involvement of top management in any effective improvement process.

Breaking down interdepartmental barriers is an essential requirement for better teamwork and business process improvement. New integrative mechanisms such as matrix structures and project teams can only flourish where these barriers are removed. Improved business processes, such as simultaneous or concurrent engineering, can lead to significant benefits in the reduction of product development cycle times. The lead times required to get from the initial concept of a new product through the various stages of design, engineering, testing, manufacturing and marketing, can also be reduced significantly, thus producing considerable savings as well as enabling the company to be ahead in the marketplace. These kinds of issues cannot be tackled by Quality Circles.

Another criticism of Quality Circles is that they are voluntary and thus may exclude many people in organizations who should be involved because their expertise or insight is necessary for effective problem solving, or because TQM is a philosophy which by definition involves everyone. The proposition that Quality Circles only involve volunteers is in itself a questionable one. Whilst in Japan this has been the custom and practice, there are arguments in favour of what Pat Townsend from the Paul Revere Life Insurance Company in the United States has called 'non-voluntary' but nevertheless, successful, Quality Circles [1, p.76].

Quality Circles, in order to function effectively, as we have seen require certain preconditions for success. Tom Peters in *Thriving on Chaos* outlines the findings of one recent assessment of the problems of Quality Circles throughout Martin Marietta. These were [1, pp.300–301]:

- misunderstanding of the concept and process by upper and middle management, creating false expectations
- resistance to the concept and process by middle managers and supervisors, often verging on outright sabotage
- empire-building by the quality circle office, substituting the illusion of immediate success for the long-term goal of institutionalizing the quality circle process
- poor and 'one-shot' training for circle members, supervisor-leaders, and managers
- failure to prepare the organization to provide incentives for participation in quality circles
- failure of the organization to implement circle proposals
- failure of the organization to measure the impact of quality circle participation – on defect rates, productivity rates, attrition rates, accident rates, scrap rates, grievance rates, lost-time rates, and so on
- failure to develop and codify a set of process rules prior to forming the first circles
- moving too fast – forming more circles than the quality circle office or the organization can deal with adequately

Our conclusion about employing the Quality Circle or problem-solving process alone, as a means of implementing TQM is, therefore, that it is too restrictive and limited if the full benefits of TQM are to be harvested. The evidence is that there are considerable benefits to be obtained both in terms of changing the organizations' approach to involving people and in terms of cost savings, but the approach is only a partial one fraught with dangers if not carefully managed.

3.6 STANDARD IMPLEMENTATION METHODOLOGY OR 'PACKAGED' APPROACHES TO IMPLEMENTATION OF TQM

It is not surprising, given the pre-eminence of the ideas of Deming, Juran and Crosby, that many companies have chosen to adopt the philosophy and implementation methodology of one of these well known advocates of TQM. From an individual company's point of view it may seem that the easiest and simplest way of getting started is to have a consultant representing one of these well known approaches and follow the advice given. According to Tom Peters in order to be successful it is important to have a quality system or ideology. He holds back from recommending a particular system, recognizing that it is a controversial issue. What he does emphasize, however, is the need to 'pick a system and implement it religiously', even if it is one invented within the company. He strongly advises not to begin with what he calls a 'Chinese menu' and maintains that it makes little difference which system is chosen so long as it is thorough and followed rigorously. Peters asserts that 'Most Quality programmes fail for one of two reasons: they have system without passion or passion without system. You must have both' [1, p.74]

What then are the elements of these different systems? Let us examine Deming, Juran and Crosby in turn.

3.6.1 W Edwards Deming's philosophy

Earlier, in Chapter 2, we outlined Deming's 'fourteen points' which he saw as the basis for bringing about a transformation in American industry. These fourteen points however, have been misinterpreted by some managers as steps to be followed sequentially as if they were an action plan. This is not the case. Clearly Deming intended his fourteen points to be adopted but he does not provide a prescription as to how to go about it. Likewise with the seven deadly diseases, he expects management to eradicate these but he does not say how this should be done. When he wrote his book *Out of the Crisis* outlining these ideas he anticipated that it would take between 10 and 30 years for American organizations to achieve a competitive position. To help managers implement his ideas he provided a list of 66 questions, many of which are sub-divided, and the answers to which, managers would need to find for themselves. To illustrate the nature of these questions the following example is given [7, p.157]:

6a Why is transformation of management necessary for survival?
b Are you creating a critical mass of people to help you to change?
c Why is this critical mass necessary?
d Do all levels of your management take part in the new philosophy?
e Can any of them initiate proposals for consideration? Do they?

The fourteen points can thus be seen as producing guidelines but they are not steps in an action plan. We are not saying, however, that Deming does not offer an action plan. This is outlined by Professor Tony Bendell in the Department of Trade and Industry's booklet *The Quality Gurus*'. The plan contains the following seven points[8]:

Deming's action plan

1. Management struggles over the 14 points, deadly diseases and obstacles and agrees meaning and plans direction.
2. Management takes pride and develops courage for the new direction.
3. Management explains to the people in the company why change is necessary.
4. Divide every company activity into stages, identifying the customer of each stage as the next stage. Continual improvement of methods should take place at each stage and stages should work together towards quality.
5. Start as soon and as quickly as possible to construct an organization to guide continual quality improvement.
6. Ensure everyone takes part in a team to improve the input and output of any stage.
7. Embark on construction of organization for quality.

This action plan appears to be a logical sequence of activities for managers to adopt. On its own, however, it is unlikely to be sufficient. Managers are still likely to need help to interpret these guidelines and to provide them with more specific answers to the question 'What do we do?'. To overcome this problem, the Deming Association has a register of consultants who are advocates of the Deming approach who provide training courses for managers. These help them to acquire a better understanding of the underlying principles and of the necessary steps required to implement the Deming philosophy. Because Deming places a great deal of emphasis upon the analysis and removal of variance, the implementation of his approach entails the application of Statistical Process Control (SPC) techniques[9]. The application of SPC is often seen to be only relevant to certain kinds of manufacturing activities and hence inappropriate for non-manufacturing and service-based organizations. Advocates of the use of SPC argue that any process can be analysed using the technique, providing suitable parameters to measure are selected. It is this reliance on the application of SPC, which is integral to Deming's approach, which has caused some organizations to reject it altogether. SPC, however, is merely a tool to be used within a much broader framework of ideas, all of which form a coherent philosophy. Rejecting the rest of Deming's ideas because of a dislike of the use of statistical techniques is equally as unwise as the adoption of statistical techniques in a vacuum, which is what some companies appear to have done.

3.6.2 The Juran approach to implementation of TQM

This standard implementation methodology is usually promoted through the Juran Institute or through agencies for the Juran Institute. Converts to the Juran approach are able to purchase a set of video cassette tapes for use in training sessions in their own companies to institute and maintain a quality improvement process. The package contains the following tapes [10, Section 11, p. 17]:

- Proof of the need
- Project identification
- Projects to improve product saleability
- Organizing for improvement
- Organizing for diagnosis: The diagnostic journey
- Operator controllable errors
- The diagnostic tools
- The remedial journey
- Motivation for quality
- Holding the gains

Implicit in the set of topics is a logical sequence of activities for companies to employ in order to implement the TQM process. But Juran has set out independently from these topics his famous ten steps:

The ten steps to Quality Improvement

These are [10, Section 22, p.6]:

1. Build awareness of the need and opportunity for improvement.
2. Set goals for improvement
3. Organize to reach the goals (establish a quality council, identify problems, select projects, appoint teams, designate facilitators)
4. Provide training
5. Carry out projects to solve problems
6. Report progress
7. Give recognition
8. Communicate results
9. Keep score
10. Maintain momentum by making annual improvement part of the regular systems and processes of the company

Managing the implementation process

How should the implementation process be managed?

We have already noted that Juran paid a great deal of attention to the management of quality, right from his early days. Recently in an interview he was asked whether or not there is a basic pattern which companies which have successfully introduced TQM, have followed. His observations reflect his earlier views about how the implementation process should be managed. This was his reply[11]:

> Companies that have implemented successful quality management programmes have done so by taking some giant steps, usually in the following order:
>
> - They undertook project-by-project improvement at a revolutionary pace. Those improvements provided impressive gains in product performance and in waste reduction, while arming their managers with experience and expertise in the improvement process.
> - They then undertook to improve their quality-planning process in order to close off the hatchery that was creating their chronic quality problems.
> - Finally, they established strategic quality management by opening up the business plan to include planning for quality.

Strategic Quality Management he defined as 'a systematic approach for setting and meeting quality goals throughout the company'[10]. It represents a culture that is practised at the highest levels of the organization and for its introduction it requires an initiative by senior management supported by personnel participation.

Section 6 of Juran's *Quality Control Handbook* expands upon the subject of Company Wide Quality Management. He suggests that companies should [10, Section 6]:

- Establish broad business goals
- Determine the deeds needed to meet the goals
- Organize – assign clear responsibilities for doing those deeds
- Provide the resources needed to meet those responsibilities
- Provide the needed training
- Establish the means to evaluate actual performance against goals
- Establish a reward system which relates rewards to performance

Training which is mentioned above is a feature of Juran's approach upon which he places a great deal of emphasis. This for him is an essential ingredient and not only requires a break with tradition but also involves a change in the culture of the organization. Traditionally training for quality has been concentrated in the quality department but Juran maintains that the 'real need is to extend such training to the entire management team involving all functions and all levels of management' [11, p.10]. Moreover he argues strongly that senior managers should be the first to undertake the training for quality. His reasons for this are that[11]:

- By being first, senior executives become better equipped to review

proposals made for training the rest of the workforce.
- By setting an example, the senior executives change an element of the corporate culture; that is, for lower levels to take the training is to emulate what has already been done at more senior levels within the organization.

Juran also says something worth noting about the importance of senior management's role in establishing a broad-based task force to develop the company's approach to training for quality. Such a task force should[12]:

- Identify the company's need for training in managing for quality
- Prepare a curriculum of courses that can meet these needs
- Identify which categories of personnel should take which bodies of training
- Identify sources of needed training materials whether self developed or acquired from suppliers
- Identify the need for leaders, i.e. trainers and facilitators
- Propose a time table
- Estimate the budget

Training plans developed by a company task force are more likely to meet the needs of the company than otherwise – even though it may take longer.

Commenting on the need for training to start at the top of the organization, he maintains as far as senior management are concerned 'Their instinctive belief is that upper managers already know what needs to be done, and that training is for others – the workforce, the supervision, the engineers. It is time for a re-examination of this belief[12].

It is noticeable in each of the above sets of recommendations no mention is made of the customer. Juran has remedied this, however, in his book *Juran on Planning for Quality*. In this he outlines the steps of a *Quality Planning Road Map*. This consists of the following nine steps[12]:

1. Identify who are the customers
2. Determine the needs of those customers
3. Translate those needs into our language
4. Develop a product that can respond to those needs
5. Optimize the product features so as to meet our needs as well as customer needs
6. Develop a process which is able to produce the product
7. Optimize the process
8. Prove that the process can produce the product under operating conditions
9. Transfer the process to operation.

Taking these ideas together with what we said earlier about Juran's concepts of 'internal' as well as 'external' customers we can see that Juran places considerable emphasis on the importance of being continuously aware of customer needs throughout all functions of the organization as well as of those outside.

3.6.3 Philip Crosby's approach to implementation

Crosby's approach to implementing TQM is probably the most widely adopted in the west of all of the standard implementation or 'packaged' approaches. His four absolutes we have already discussed in Chapter 2. In addition he has recommended the following 14 steps for managing quality [13, p. 99]:

Crosby's 14 steps for managing quality

Description	Purpose
1. Management Commitment	To obtain commitment from all employees by management leading by example
2. Quality Improvement Team	Organizing for quality, quality improvement process planning and administration
3. Measurement	Measuring for quality on a common basis
4. Cost of Quality	To define the management tools to assess the costs of POC and PONC and reporting methods
5. Quality Awareness	Communications to employees of progress, results etc. Meetings and measurements.
6. Corrective Action	To develop formal systems for the elimination of problems
7. Planning for Zero Defects	Ad-hoc committee to help develop ideas for Zero Defects Day and consider timing of event (or alternative means of demonstrating commitment).
8. Employee Education	Quality education for all employees to secure common understanding of quality and on-going improvement.
9. Zero Defects Day	Making commitment visible.
10. Goal Setting	Helping department to establish achievable and measurable goals.
11. Error-Cause Removal	Problem identification and communication. Promote error-friendly environment.

12. Recognition	To ensure credit is given where due to raise motivation and commitment
13. Quality Council	To audit the quality improvement process in action (and feed back reactions and feelings from the receiving end).
14. Do It Over Again	To set up a new team to repeat the process.

Apart from the above 14 points, some of Crosby's other central beliefs are that Quality Circles are ineffective because, as we have discussed earlier, they are voluntary whereas total quality improvement must involve everyone. He also maintains that both suppliers and buyers should be rated for quality since their role is paramount in achieving Total Quality and that combinations of single and multiple source supply arrangements should be used where appropriate.

3.6.4 Similarities and differences between Deming, Juran and Crosby

Having reviewed the main contributions of these three best known thinkers in TQM, it is apparent that many of their ideas are similar to each other. However, each of them, as we have already seen, takes issue with the others over certain concepts and ideas. The common elements between them and the differences are listed below:

Common elements

1. Constant high levels of training and education
2. Create awareness of opportunity and constant search for improvement – permanence of the process
3. Error-friendly problem-solving environment
4. Prevention orientation and attention to detail
5. Use of self-defined measurement by all employees
6. Control of suppliers by SPC or auditing
7. High levels of non-financial recognition of employees
8. Open communication of results of projects and business performance
9. Concept of 'internal customer' and management of processes

Differences

1. Use of Quality Circles – Crosby
2. Zero defects as a performance standard – Crosby
3. Hold a zero defects day to celebrate commitment to quality – Crosby
4. Single sourcing of suppliers – Deming

5. Eliminate management by objectives and pay linked to output – Deming
6. Extensive market research – Deming
7. Use of cross functional action teams to attack problems on a project by project basis – Juran
8. Individual goal setting – Crosby and Juran

3.6.5 Other standard implementation approaches

Having outlined the major elements of the implementation approaches of Deming, Juran and Crosby, we should point out that these are not the only standard implementation or 'packaged' approaches. Many consultancy organizations offer their own ready-made training manuals lifted off the shelf awaiting a transplant into any client organization wanting an easy-to-follow solution to the building of a Quality Culture. Frequently these approaches are sold to organizations as a panacea for success and managers are expected to slavishly follow the routines laid down. Regardless of a company's circumstances, its previous experience, its state of readiness and so on, these packaged methods are adopted by senior management who perhaps have no awareness of their suitability or otherwise. Managers are then squeezed into a strait-jacket approach with which they not only feel uncomfortable, but to which they have no commitment or ownership. If the approach does not fit the culture it is almost certain to fail. Our view is that every company is different and needs to tailor its approach to suit its own situation. We are not advocating Tom Peters' 'Chinese menu' but the design and planning of a coherent set of concepts and ideas which hang together and which feel right for those managers and company employees who have to live with them. We call this approach a tailored or planned change approach.

3.7 THE TAILOR-MADE OR PLANNED CHANGE APPROACH

It may seem trite to say it, but our experience has taught us that every organization is different. Organizations have characteristics and cultures of their own which are the product of a unique combination of different variables. In the same way that people go through stages in their development from infancy through adolescence to adulthood, so too do organizations go through stages in their development. At each stage the problems faced, the structures and processes adopted, the resources available, the styles of managerial behaviour and so on, all vary. We can also confidently assert from our experience that the need for organizations to change varies alongside their individual capacities to change.

In the light of the enormous diversity of circumstances surrounding each organizational situation, it seems to us that any strategy for effectively bringing about a desired change in the way the company operates has to be tailor-made to take account of these diverse circumstances. What works in one organization may not work in another and what works in one part of an organization may not work in another part of the same organization.

Organizational development practitioners have long since recognized this and accordingly have evolved strategies for managing the change process suited to the needs of the organization with which they are dealing. In a sense, this approach to implementing TQM is nothing more than an attempt to practise what the philosophy preaches. That is, it involves identifying who the customers are, finding out what their needs are and then designing an approach to suit those needs. It also fundamentally involves the customer in the design process itself. It does not set out to produce a 'packaged' solution to the implementation process which is then sold to the customer. Another way of putting it is to say the tailor-made or planned change approach is customer led not producer led.

3.7.1 The 17 assumptions underlying TQM

From our experience and observations, we have concluded that there are certain assumptions underlying the tailor made or planned change approach to TQM. These are:

1. Total Quality Management is about improvement
2. Improvement requires change
3. Change represents a disturbance to the status quo
4. The status quo is maintained by a balance of forces or variables
5. Change or improvement, therefore, requires an alteration of these forces or variables
6. Effective change requires managers acting as change agents to alter these forces or variables in such a way that people affected own the changes and are committed to them
7. The change process thus requires consensus between all concerned
8. Consensus is achieved through involvement and participation of employees at all levels in the identification of the variables that need to be changed and in the formulation of an approach to change these forces or variables
9. Any change programme involves a set of core values
10. The core values of effective TQM strategies are those made explicit in what McGregor calls theory Y (see below)
11. Total Quality Management must affect the culture of the organization as well as the behaviour of individuals
12. Education and training may assist the process of culture change but is not sufficient by itself. Changes in the organization's policies,

structures and technical systems are also likely to be necessary

13. For change to occur there has to be a desire for change
14. Where desire for change does not exist, the agents of change will need to analyse why and create the desire
15. Resistance to change is not easily overcome. Approaches to overcoming resistance will need to be devised
16. Changing organizational culture and individual behaviour takes time – probably at least 4–5 years, possibly even 20 years and more
17. Most managers in the prevailing culture work on short-term time perspectives. Consequently few managers are able to consider providing continuing support for a project that does not produce tangible results in the short term – say 6 months. To cope with this paradox an appropriate strategy will need to be employed.

3.7.2 McGregor's theory X and Y assumptions

In his book *The Human Side of Enterprise* Douglas McGregor discusses assumptions about human behaviour in terms of theory X and theory Y.[14] These are shown in Figure 3.4.

These assumptions, as McGregor recognizes, are not finally validated. Nevertheless, he agrees that they are far more consistent with existing knowledge of the Social Sciences than those of theory 'X' and, on the surface, do not seem particularly difficult to accept.

As we said above, the TQM approach has a set of core values of which 'theory Y' is a part.

3.7.3 Reconciling short-term and long-term perspectives – a dual track approach

To address the paradox outlined above – that is that TQM is a long-term strategy to bring about behaviour and cultural change, but managers need to produce tangible results in the short term – we recommend a dual track approach to installing a Total Quality Process. The following section is an example of a tailor-made proposal for installing a TQM process into a small company, and illustrates how the conflict can be resolved.

3.8 THE ROLE OF AWARDS IN IMPLEMENTING TQM

Many organizations have taken steps to implement Total Quality Management processes using the criteria developed for selecting Quality Award winners. Quality Awards are a means by which companies can achieve recognition for their efforts and these have become increasingly popular in recent years.

Theory 'X'

The first set of assumptions underlying the traditional approach to managerial decision making and action is:

1. The average human being has an inherent dislike of work and will avoid it if he can.
2. Because of this human characteristic of disliking work, most people must be coerced, controlled, directed, threatened with punishment, to get them to put forth adequate effort toward the achievement of organisational objectives.
3. The average human being prefers to be directed, wishes to avoid responsibility, has little ambition, wants security above all.

These assumptions have had a profound influence upon the development of managerial principles and hence managerial practice up to the present day. As McGregor points out – they would not have persisted if there were not a considerable body of evidence to support them. Nevertheless, there are many readily observable phenomena in industry and elsewhere which are not consistent with this view of human nature.

Theory Y

The second set of assumptions derive from an accumulation of knowledge about human behaviour in many specialised fields. These assumptions, referred to as Theory 'Y', are:

1. The expenditure of physical and mental effort in work is as natural as play or rest. The average person does not naturally dislike work.
2. External control and the threat of punishment are not the only means for bringing about effort towards organisational objectives. Man will exercise self direction and self control in the service of objectives to which he is committed.
3. Commitment to objectives is a factor of the rewards associated with their achievement. The satisfaction of needs can be seen as rewards.
4. The average human being learns under proper conditions not only to accept but to seek responsibility. Avoidance of responsibility, lack of ambition and emphasis on security, are generally consequences of experience – not inherent human characteristics.
5. The capacity to exercise a relatively high degree of imagination, ingenuity and creativity in the solution of organisational problems is widely, not narrowly, distributed in the population.
6. Under the conditions of modern industrial life, the intellectual potentialities of the average human being are only partially utilized.

Figure 3.4 McGregor's theory X and theory Y.

Best known of the quality awards are probably the Deming Prize based in Japan but with a version available internationally, the Baldrige Award in the USA, and the European Quality Award. In the UK the British Quality Association (BQA) have also recently launched a British Quality Award based upon the same criteria as the European Award. There is a separate award in Northern Ireland which uses a different model.

3.8.1 The Deming Prize

The Deming Prize was the forerunner of the national quality award fashion and was established by the Japanese Union of Scientists and Engineers. It is a recognition of the impact of the teachings of the late Dr W. Edwards Deming on Japanese economic performance through quality improvement. On average there are around ten Japanese winners each year. Its international version, the Deming International Award, was established in 1987 and has only rarely been won.

The award uses the following criteria to assess organizations:

- Understanding and zeal of top management for quality
- Policy, goal and purpose (i.e. an examination of the philosophy – such as Deming's 14 points – being used by the organization to drive its quality improvement efforts)
- Organization for quality, including systems and personnel involved
- Education in quality, incorporating its robustness in statistical principles and methods of teaching and its extension to suppliers
- Implementation of the quality improvement process
- Future policy, plan and measures of performance. This criterion deals with the organization's response to change

3.8.2 The Baldrige Award

The Malcolm Baldridge Award, named after the late Secretary of Commerce, was established by an Act of Congress in 1987. Criticized by some for being over-bureaucratic and tainted to some extent by the later crashes of some of its winners (most notably the Wallace Corporation who went to the wall less than 12 months after winning the award), the award nevertheless has generated increasing interest each year.

The drop-out rate for the award is, however, quite high with only around 10% making it past the first stage. Over the years it has not found sufficient world class entrants to award all the possible prizes, with service sector awards being particularly thin on the ground. From a theoretical point of view the award criteria are more closely related to the ideas of Joseph Juran than they are to Deming. The award examiners are given a range of indicators of Total Quality performance to look for. These are set out below:

- Leadership – clearly defined quality values, goals and plans for their achievement and communication
- Information and Analysis – its reliability, currency and usability
- Strategic Quality Planning – level of leadership involvement in and integration of quality plans into the business
- Human Resource Utilization – developing the full potential of the workforce
- Quality Assurance processes for products and services
- Quality Results – how improvement and quality performance is measured
- Customer Satisfaction – including the customer's role in defining the requirements for quality

3.8.3 The European and British models

These are dealt with together since they work from an identical model which is reproduced in Figure 3.5. The award was launched by the European Foundation for Quality Management in 1991 with Rank Xerox the first winner in the following year. Rank had set itself two ambitious goals:

1. 100% customer satisfaction
2. 100% employee satisfaction

These are reflected in the European Quality Award criteria themselves.

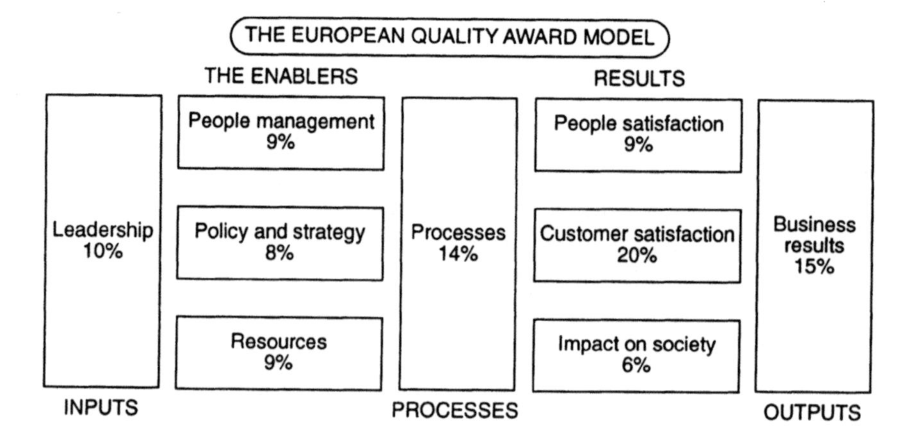

Figure 3.5 The European Quality Award criteria.

3.8.4 What's in a badge?

So what do the companies get who seek out these awards? From a marketing perspective the award gives the company a high profile quality image which carries far more weight than a standard. They get the opportunity, especially so with the EFQM award, to research best practice among other award applicants – a good route to competitive benchmarking. In return the company must share the secrets of its success with others and be prepared to publicize the award.

3.8.5 The downside

If there is a downside, it has to be the extensive documentary evidence and assessments required for these awards which some would argue would be time better spent in improving service. Indeed, Deming himself was never a great fan of his award – in some ways his analogy of such award systems to lotteries is apt. However, most award applicants, even the unsuccessful ones, admit that the work required acted as a catalyst to their improvement efforts and highlighted important areas of their business requiring attention.

From a nationalistic perspective we have some other worries about the value of awards in raising standards. Recently a Swiss-based management research forum produced its *World Competitiveness* criteria which provide an independent summation of several key economic and industrial indicators governing countries' competitiveness. For many years the Japanese have topped this list, only occasionally being knocked off by the Swiss or the Germans. Britain still languishes in the bottom half of the table and this year is two places behind Ireland (18th and 16th respectively) – you remember Ireland, the country comedians always liked to laugh at.

Our experience as consultants is that many companies embarking on quality improvement processes do not possess the necessary will or management aptitude to make it happen, and unfortunately quality awards at a national level just seem to be passed around the premier league sides like Rank Xerox, Milliken, Rover, Motorola, etc.

The current obsession with awards and assessment is only likely to further highlight the performance of the excellent few whilst leaving the mediocre many continuing to provide the often very poor level of customer service we all experience daily.

REFERENCES

[1] Peters, Tom (1987) *Thriving on Chaos*, Pan Books in association with McMillan, London

[2] Pike, R John (1992) Gates Hydraulics Limited, 'Total Quality Management

and the Single Market'. In *Cases in European Business* (ed Jill Preston) Pitman, London, pp. 129–130

[3] Barra, Ralph J (1989) *Putting Quality Circles to Work*, McGraw Hill, New York

[4] Dolan, Patrick (1990) 'Quality Circles', Department of Trade and Industry, London

[5] Robson, Mike (1988) *Quality Circles A Practical Guide*, 2nd edition, Gower, Aldershot

[6] Robson, Mike (1988) *Quality Circles*, 2nd edition, Gower, Aldershot, p. 17

[7] Deming, W Edwards (1982) *Out of the Crisis*, Cambridge University Press, Cambridge

[8] Bendell, Tony, The Quality Gurus, Department of Trade and Industry, London

[9] Mortimer, J (Ed) *Statistical Process Control*, IFS Publications, Kempston, UK

[10] Juran, J M and Gryna, F M (1988) *Quality Control Handbook*, 4th edition, McGraw Hill, New York

[11] Juran, J Quality Management Programmes. Les Picket talks to Dr Joseph Juran, *Training & Development Journal*, August 1992

[12] Juran, J (1988) *Juran on Planning for Quality*, Free Press, New York

[13] Crosby, P B (1984) Quality Without Tears, McGraw Hill, New York

[14] McGregor, Douglas (1960) *The Human Side of Enterprise*, McGraw Hill, New York

PART TWO

A Tailored Approach to Implementation

The Seven-P process: A practical approach

4

4.1 AN INTRODUCTION TO THE SEVEN-P PROCESS

There are probably as many systems for implementing a Total Quality Management process as there are people and organizations selling their particular approach. Deming's 14 points, Crosby's 14 steps, Juran's 10 steps, Conway's 6 tools. At the risk of antagonizing our readers with yet another approach, let us introduce you to what we term the Seven-P Process. It differs in one important respect from the preceding recipes in that it is chronological. This is in answer to the needs of many of our clients over the years who have found difficulty in planning their implementation process with several overlapping and integrated steps. One of the most common cries for help we receive is 'OK so we understand the 14 steps/points or whatever, but what do we do first and when do we start step 2, 8, 13 etc.?'

The aim of the Seven-P process is to give TQM implementation teams a clearer vision of the path they need to follow in putting into practice any of the main systems offered by the quality gurus. The Seven Ps can be overlaid on Deming, Juran and Crosby's approaches and are applicable no matter whose particular philosophy you choose to adopt. There are many examples of organizations who have succeeded using Crosby, as there are with Deming. There are probably far more however, from whatever camp, who have come seriously unstuck through lack of direction.

4.2 THE SEVEN PS IN OUTLINE

The Seven Ps are:-

1. **Positive commitment**
2. **Planning**

3. **P**articipation
4. **P**rocess control
5. **P**roblem identification
6. **P**roblem elimination
7. **P**ermanence

In the next four chapters we will be setting out the detail behind each of the 7 Ps under three stages – getting started, getting organized, managing the continuous improvement process, and making it stick.

In each chapter we will be reviewing the essential steps and the **nice to do** things using examples from clients and other companies covered in our research and direct consultancy support. Each chapter concludes with a checklist of ideas, questions and issues which should be addressed in passing through that phase of the implementation process. For the steering team with a mental block on what to do next, we offer these as an unstructured suggestion kit for keeping up the momentum.

Even before we get to the first P there are a number of preliminaries (perhaps we should have had 8 Ps?) any organization needs to address. These cover the need for change, organization culture and climate. Culture is distinguished from climate here in that the culture is seen as a largely fixed set of behaviours whereas climate represents the current mood or preoccupation of the people in the organization. A recent redundancy exercise will affect the climate but not necessarily the culture. TQM must affect both if it is to take hold.

4.3 THE NEED FOR CHANGE

Chapter 1 discussed some of the reasons why there is a need for a new approach to Quality Management and why it has become such an important issue for organizations in the 1990s and beyond. It is important for anyone embarking on TQM to understand why they are doing it. If the only reason is that the concept is appealing (who can be against getting it right and improving?) or that your customers are doing it, stop now. You are heading for the fad pit.

The fad pit is well known. Managers at the top or new CEOs coming in with their new brooms latch onto the latest piece of research, dressed up in a catchy bit of jargon, to emerge from Harvard (they probably did their MBA there). This then becomes the panacea for all the organization's ills. Often aided by the gullible human resource wallah, management rush about, committing large amounts of energy, money and hot air on exhorting everyone to try this new remedy for themselves. This is greeted with a mixture of scepticism and polite compliance from those longer in the tooth. These people then suddenly transform from being useful employees to become 'the people we need to get rid of anyway' and are variously described as 'playing into our hands' or 'signing their own P45s' when they continue to refuse to join the zero defects day or quality party.

These people are not barriers to change – they are hedgehogs. When faced with a threat they curl up and stick their spines out. There are two things to do with a hedgehog in a posture like this. One is to run it over – the 'playing into our hands response'. The other is to leave it alone and get on with your own job in resolute and determined fashion. The hedgehog has every right to curl up – he **has** seen it all before – why should he believe you this time and like as not he **was** 'doing quality' before you could even spell the word. The CEO and top team who can meet the first of the Seven Ps will understand that they simply have to be consistent (unlike Worzel Gummidge, you cannot 'jus'pu' on moy kwolii ed now') and insistent over a long period in order to persuade the hedgehogs to uncurl. But uncurl they do – we have seen some remarkable 'Pauls on their way to Damascus' in our years of introducing TQM. Don't be disheartened if it takes time to convert people to these concepts – many of them will wonder why you think they weren't already committed to quality anyway. One of the chief tasks of those leading any organization is to keep repeating the same messages to people in different ways at every opportunity. This is not as easy as it sounds. The trick is to be consistent and to be able to live and breathe the messages you are trying to get others to believe. Chapter 5 covers this in more detail.

Unfortunately, in the fad pit, the managers leading the process give up because they didn't achieve instant success and start to play down the process and even join in the ridiculing and winking at the jargon. They simply weren't thick skinned enough – the animal equivalent of a good TQM leader is the rhino – he does believe there are results to be gained eventually by banging his head against a brick wall, but he doesn't waste his time looking for small prickly creatures to annihilate.

Another symptom of the fad pit is experienced when management fails to appreciate just how fundamental a change in culture is required in TQM. Management sees TQM as a cost-cutting exercise, an extension of ISO9000, a way to motivate people or a way to bypass a troublesome trade union. Strategies based on these single track issues are doomed to be short lived and probably will reinforce already apathetic views of management's commitment to quality.

Introducing a Total Quality strategy requires attention to a number of key issues; we like to describe them as a three-legged stool – the analogy being that if you fail to pay attention to any one leg of the stool, the whole structure becomes unstable. The three legs of the stool are shown in Figure 4.1.

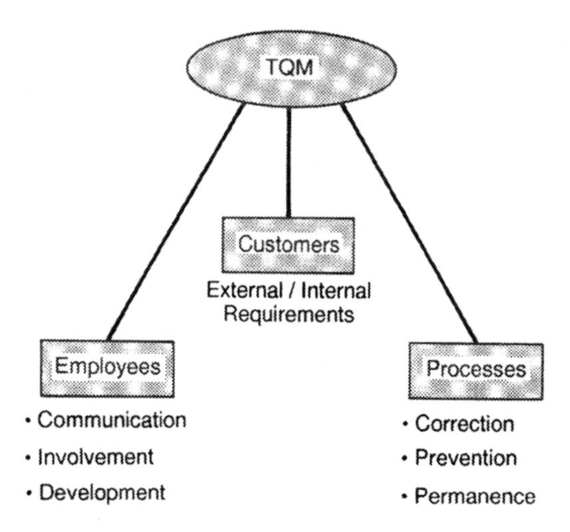

Figure 4.1 TQM implementation strategy: key issues. Like the stool, stability requires constant attention to all three legs.

4.3.1 Customers

First among equals is the customer leg; not just the people who pay your wages, but also the internal customer. Managing the internal customer/supply chain as well as the external one is essential if you are to produce the highest quality of conformance at the lowest cost. Chapter 7 deals in more detail with this matter under the subject of **P**rocess Control. It is surprising how few companies really take this concept on board seriously, considering that Total Quality is essentially about the customer. Unfortunately Quality Assurance approaches such as ISO9000 make hardly any mention of the customer and focus attention almost entirely on internal matters related to product or service delivery. This tends to blind companies into thinking that the way to improve quality is through better procedures and more discipline.

Customer surveys

A good way to start any Total Quality initiative is to survey your customers. This doesn't have to be a complex issue – although it is often useful to get outside advice on structuring the survey in order to ensure objectivity and statistical correctness. Questions should address both the product and service issues. Recent studies have shown that the proportion of total product cost attributable to internal services has risen from 50% in the 1960s to 70% in the 1980s[1]. Examinations of customer complaints often reveal a large proportion originating in service areas rather than in the product itself. For example a recent personal experience with a mail order stationery company gave rise to over 10 different complaints or failures on the supplier's part. These included losing the order, computer input error, failing to record payment, wrong quantities and even mailing an unwarranted repeat invoice in two separate envelopes – one for each page. None of these failures had anything to do with the quality of the goods ordered; in fact the supplier probably considers itself as providing quality goods!

The form of the customer survey does not have to follow the traditional questionnaire format – personal interviews through telephone, focus groups, customer days, sales rep discussions and feedback all provide more information, but beware of the **friendly factor**. People will often not raise minor irritating problems on a personal basis but are quite prepared to supply these through a more anonymous means such as a questionnaire. Hotels and some restaurants have used surveys for many years as a means of extracting data on such matters. It is often not the single big mistake that switches a customer off your organization but a gradual build up of niggles; if you can't count the straws on the camel's back, you'll never see the last one coming. If you don't believe this, ask yourself whether you would take a light bulb back to the vendor if it broke after 2 hours of use. Most of us would not, but if it happened twice we would probably switch brands or vendors or both. Many organizations who think they have low customer complaints simply haven't asked the customer. British Rail would have to multiply its complaints by a factor of anywhere between 50 and 200 if **all** customers complained about **all** service failures. As a rule of thumb, only 4 in 100 customers with a bad experience bother to complain to the provider. However, each unhappy customer probably tells 10 of his friends, relatives, acquaintances and fellow sufferers about the incident. This is shown in Figure 4.2.

The effect of this hidden drain on future business is analogous to the release of a virus into the population. Not only is the recipient of the virus (problem) infected, he infects others, who in turn infect others. Each infected party builds up an antibody to your organization and consequently never becomes a customer in the first place. The easiest way to stop customer complaints is to have no customers!

A number of questions need to be answered before designing the survey. Firstly, who are your customers? Are they just existing users or all potential users in the population. The point of contact is also important; if your customers are other organizations, you may get a different response from an executive than from the hands-on customer who perceives service quality more keenly. If you supply photocopiers, do you ask the buyer, the CEO or his secretary?

Secondly, what do you hope to achieve from the survey? Are you simply testing the water for comparative purposes or seeking specific information

The Iceberg

Only 4 in 100 disappointed customers complain
Each unhappy customer tells 10 others of their experience

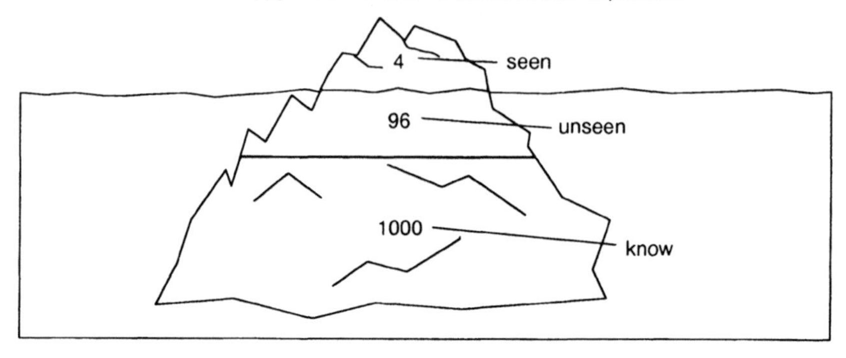

The Implications

Only 1 in 10 disappointed customers will come back
It costs 5 times more to attract a new customer than to retain an existing one

Figure 4.2 Customer complaints iceberg showing relevance of complaints received to actual impact on business of disappointed customers.

about certain areas of your business? This question is also relevant to the first. If you are seeking information, go to the point of contact with the product or service (in the case of the photocopier, the secretary). However, if you are trying to influence future business, the end user is less likely to have knowledge of alternatives or be able to influence buying decisions. Construction of questions needs to have the knowledge of the interviewee in mind.

Thirdly, do you intend to repeat the survey? If so, at what intervals? Surveys are valuable sources of data if used in a programmed and structured manner. Once off efforts are of less value.

Fourthly, what are you prepared to do with the results? If you receive a negative or unfavourable response to some aspect of your product or service, or suggestions for improvement, are you prepared to do something about them? It is important to remember that a survey is an intervention in the marketplace. It will not happen in isolation and will have unseen effects. Customers may imagine it is a precursor to a change in product or service or an intention to improve performance in a particular area. As a result, expectations may be raised or the survey may act as a catalyst to future complaints. Do not be surprised if the frequency of complaints rises after a survey. If you show interest in your customers' views, they are more likely to tell you about them. Or, put another way, sticking your head out of the trench is an invitation to get shot at! But it may be the only way to win.

Question design

Most issues in customer perception surveys tend to be subjective in nature. This means that data coming out of the survey may have very low reliability. To overcome this potential problem, subjectivity needs to be reduced as far as possible. There are a number of ways to do this.

1. Make sure the parameters are visible to the respondent. Try to be specific about that which the customer actually experiences about the product or service. For example, rather than ask the customer to rate **responsiveness**, ask how often calls are returned within one hour.
2. Use scaling or ranking to make it easier for respondents to place their views in an objective framework. Consider also ease of transfer of the data to a computer.
3. What are you actually delivering to your customer? Be specific and avoid generalities such as **service** or **quality**. Define the key variables in the service or product and then view them from the users' perspective.
4. Avoid asking for comparative information (e.g. how do we compare to our competition?) if the respondent is likely to have only limited knowledge of alternative suppliers.
5. Avoid complex questions and ask only one at a time. Do not couch secondary questions in the main one.
6. Avoid odd numbers of boxes if you wish to steer clear of the **middling tendency**, even though 5 or 7 boxes are statistically more correct in representing a normal distribution.
7. On the form itself, leave plenty of space for comments and keep the lettering large and clear with plenty of space between questions.
8. Finally, do not ask too much. The longer the survey, the less likely you are to get an accurate or complete response. A folded A3 sheet is probably the limit.

Sampling

The next step is to decide the sample size. There are established statistical principles for doing this which are not the subject of this text. Suffice to say that the sample should have sufficient breadth (percentage of population) and depth (stratas of region, industry, job-type, person-type, etc.) to allow the results to be interpreted with minimal risk.

Format

Having set the objectives of the survey and decided the information you need, there are a number of ways to obtain the results. The methods can be divided into two broad categories – mass and direct. Mass methods include telephone surveys and questionnaires. Direct methods include structured face-to-face interviews, focus groups, sales returns and complaint analysis. The last two are important. Many organizations spend huge sums of money with market research people, only to be told what was already staring them in the face for free. The value of customer complaint data, sales department and front line staff nous should not be underestimated. One client of ours simply had all his managers go back over their files for a year prior to a seminar which we ran for them and dig out any letters, forms, notes or other indicators of customer feelings they could find. This unstructured approach revealed a considerable amount of information which was highly relevant and even if not scientifically obtained, gave useful pointers to the design and target of future survey instruments.

The advantages of mass methods are the breadth of sample possible, ease of use and the ability to contract out the donkey work. Disadvantages are mainly concerned with the relatively shallow level of enquiry possible. Most people automatically turn to questionnaires for mass sampling, but the telephone survey is an alternative which ought to be considered. Although more expensive than mailing a questionnaire, the response rate is much higher (return rates for questionnaires are often as low as 1%,) and it is considerably cheaper per head than face-to-face interviews.

Questionnaires have the advantage of greater honesty in answers, particularly if anonymity is maintained, but have the drawback of tending to be weighted by those with a specific good or bad experience of the organization or by those with time on their hands to do the survey. Gross distortions can result if one is not careful.

Telephone surveys have a number of other advantages. These include:

Time – quicker contact and response periods, particularly if the exercise is well staffed and the points of contact are not too difficult to get hold of.

Customer response – often more prepared to participate (unless this is the third survey today!) as it saves them time reading and answering a questionnaire. Honesty levels are not as high as in anonymous questionnaires (although verbal assurances of anonymity can be given), but they are probably higher than in face-to-face interviews.

Interpretation – questions can be clarified or answers probed, depending on the skill and knowledge of the interviewer. Responses are usually more spontaneous and not doctored through the command chain.

The disadvantages of telephone surveys include:

Time – costs make it necessary to limit call length and thus the amount of information which can be obtained.

Customer response – there is usually no opportunity to consult others in the organization, so answers are off the top of the head. Important points, incidents or views may be missed without time to reflect on the questions.

Interpretation – the **friendly factor** is still present to some degree. Also, if the respondent is busy, he or she may simply give answers to get the interviewer off the phone as soon as possible.

Mixing mass methods with a small number of face-to-face interviews of a sample of responses is a well tried and tested approach which attempts to obtain the best of both worlds. Rank Xerox for example, have long been a user of customer surveys to support the quality initiative. A simple type of survey instrument used by British Gas is shown in Figure 4.3.

4.3.2 Processes

The second leg concerns processes. Quality requires robust, reliable systems and procedures in order for the organization to consistently sense and meet the customer's needs. This leg is where systems such as ISO9000 come into play. They underpin the improvement process by ensuring you do not lose sight of the fundamentals in your rush to introduce new ideas. Theodore Levitt at Harvard[2] identified a customer service model which reveals 4 layers of customer service from **generic** to **potential**. The principle is shown in Figure 4.4.

Levitt describes the generic as the basic product offered or advertised. For example a taxi firm provides door to door motor transport. The **expected** product includes the implied terms of the contract; e.g. that the driver takes the most direct route, is quick but careful and doesn't smoke. The augmented layer is the area of **obvious** added value, such as the vehicle comfort and trim, fast response times, help with baggage, opening the door, etc. The outer ring is rarely apparent. Famous names such as Nordstrom, Stew Leonard, Disney and their ilk are working in this outer layer. They are attempting to enhance the product to such a degree that it takes on a whole new meaning. Disney took funfairs into theme parks; Nordstrom and Leonard aim to make shopping an experience; they are ultimately selling something well beyond the generic or expected product and have taken a quantum leap past the mundane areas of added value. In the case of the taxi firm the **potential** ring might include in-car entertainment (chosen by the customer rather than the heavy metal tape played by the driver), personalized cabs, birthday cards and flowers to key clients, free outings for the kids and their friends, lunch breaks

Please tick in box

1. What is your general opinion of the service provided by your local Gas Region?	Excellent	☐	10. 0
	Very Good	☐	– 1
	Good	☐	– 2
	Satisfactory	☐	– 3
	Fair	☐	– 4
	Poor	☐	– 5
	Very Bad	☐	– 6
2. **(a)** Have you had any work done to the gas pipe (i.e., the pipe between the gas main in the street and your customer control valve) supplying your property, within the last 3 months?	Yes	☐	11. 0
	No	☐	– 1
(b) **IF YES,** What is your opinion of the most recent job done?	Excellent	☐	12. 0
	Very Good	☐	– 1
	Good	☐	– 2
	Satisfactory	☐	– 3
	Fair	☐	– 4
	Poor	☐	– 5
	Very Bad	☐	– 6

3. Do you have any other comments on the service provided by your local Gas Region?

4. Which, if any, of these things have you done in the last 3 months . . .

(a) Visited a Gas Showroom for any reason?	Yes	☐	17. 1
	No	☐	– 2
(b) Queried or sought advice about any bill or account from local Gas Region?	Yes	☐	17. 4
	No	☐	– 5

Thank you for your help. Please post this card without delay. No stamp is needed.

Figure 4.3 Extract from British Gas Customer Survey instrument.

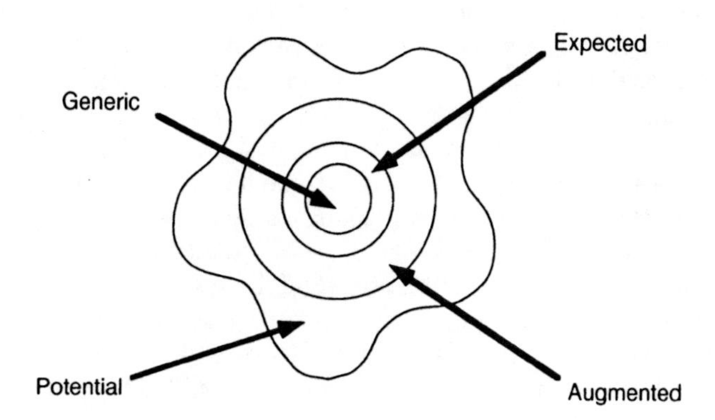

Figure 4.4 Theodore Levitt's Total Product Concept showing layers of product attributes.

on longer trips, drinks and ice cream from the in-car fridge when in a traffic jam. Most of these ideas have probably never been encountered by the average taxi firm.

Current theory is that organizations which want to stay ahead of the competition should focus attention on the outer two rings; in other words, supplying superior customer service beyond the normal boundaries of **added value** concepts. However there are many examples of people who have come unstuck with this approach through concentrating solely on the added value side without ensuring thorough application of the basics. Many of the problems of the British car industry stemmed from their rush to incorporate the latest technical and electronic wizardry without making sure of its efficacy or anticipating irritating side effects like locking you out while the engine is running! It is of little interest to the motorist seeking a service, to be feted at his local garage, provided with taxis or given a temporary car, if his own vehicle comes back with more faults than it went in with. Free ice cream and birthday cards will not compensate for the taxi which doesn't turn up when it was promised. Our advice would be to leave Levitt's outer rings until you have got your generic and expected rings under control.

The Cost of Quality

It is possible to measure the effectiveness of the process leg, utilizing an initial Cost of Quality study to highlight the need for change and the opportunities available from improving the processes in the system. We are using the word **system** here in the wider sense used by Deming to mean all the activities in the operation of the organization. These activities are described as processes in which there are inputs and outputs. Thus, a process is not confined to mainstream product/service delivery issues but to any activity in which something is converted into something else; e.g. a question into an answer, a discussion into an agreement, a performance review into a target or objective. The Cost of Quality is the result of all variations in the processes at work in the system together with the costs of controlling such variations. A variation can be defined as a departure from the target (mechanisms for controlling variation will be discussed under Process control in Chapter 7). The Cost of Quality is therefore the costs of preventing variation, inspecting or testing the system for variations and the costs of resultant planned or unplanned variations. A simpler way of thinking about this is to use Phil Crosby's definitions; the **Price of Conformance** and the **Price of Non-conformance**[3]. This gives us a way of redefining organizational costs, similar to the way in which accountants might divide costs into fixed and variable or direct and indirect. A model of the cost of quality might show the whole of an organization's expenditure as in Figure 4.5.

This is an artificial model and may in reality add up to more than 100% of the total cost of operating the business, if one includes the costs

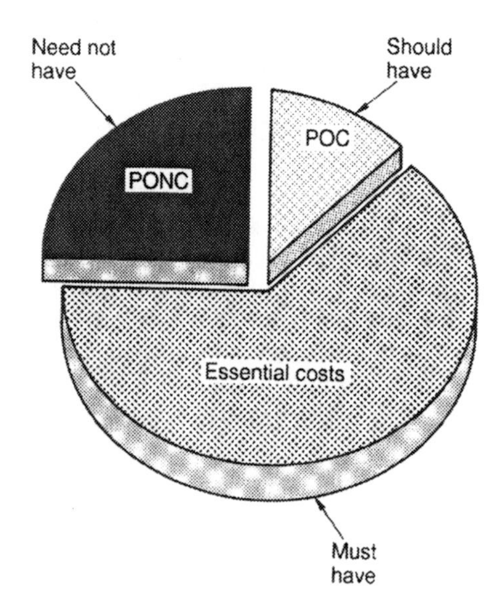

Figure 4.5 Cost of Quality showing, in Crosby's terms, expenditure of an organization split into the price of conformance (POC), price of non-conformance (PONC) and the essential costs.

of lost opportunity and lost business resulting from failures in the system, which are not directly costs to the business. Deming describes these costs as **unknown and unknowable** but in the long term they are critical to the survival of the operation. Ignore them at your peril. These costs were described earlier under the customer leg of the stool as viruses.

The **essential** costs are those associated with operating the organization with no variation. To make a ham sandwich you need a certain amount of labour, materials, energy and equipment. If any of these are missing the sandwich would not conform to its specification. In order to ensure conformance, processes of control are established such as hygiene training, preparation of work areas, calibration of slicing instruments etc. These are the Costs of Conformance. They are not essential – you might be lucky and still get conforming sandwiches but it is unlikely. Without them, you will probably incur greater variation in the form of waste, returns, complaints and ill customers. These latter costs are the Costs of Non-conformance and are the initial target of any cost of quality exercise.

Conducting the initial study

The function of the initial Cost of Quality study is clear; to establish the need for change by highlighting to top management that quality is not an added cost but an opportunity to reduce costs and improve customer loyalty. Studies over many years have shown that improving quality by reducing variation results in sometimes massive savings. Chapter 7 contains a number of impressive examples.

The study does not have to be highly accurate or complex. Margins of error of ± 20% are sufficient to prove the point. In most cases, even with a 20% reduction for error, the potential savings will still be worth going for. A useful way to carry out the initial study is to conduct a search of all reported costs and variances such as waste and scrap reports, downtime analyses, unplanned overtime, training and inspection costs. An example of an early cost of quality study is shown in Figure 4.6.

Refining the Cost of Quality system as a tool for measuring trends and targeting improvement efforts is discussed more fully in Chapter 7. The initial study simply provides a spur for action and a communication medium.

Three stages are involved in stabilizing the process leg; correction of existing faults in the system, establishing preventative systems to replace test and inspection procedures and the use of quality assurance procedures to make changes permanent. These issues are also addressed in more detail under Chapter 7 in the section on **P**rocess control.

COST OF QUALITY
COST OF NON CONFORMANCE
YEAR 1987 FIGURES UNLESS WHERE INDICATED

REASON	**£000**
SCRAP	1646
LOST TIME	1247
CUSTOMER REJECTIONS	100
OVERTIME (STAFF)	54
EXCESS STOCKS	NIL
LOST TIME ACCIDENTS	5
UNPLANNED MAINTENANCE	47
OVERDUE RECEIVABLES	477
FAULTY MATERIALS	N/A
INCORRECT ORDER INPUT	13
LATE DESPATCHES	459
TOTAL	4048

COST OF QUALITY
COSTS OF CONFORMANCE
YEAR 1987 FIGURES UNLESS WHERE INDICATED

REASON	**£000**
QUALITY CONTROL	224
TRAINING AND EDUCATION	50
PRODUCT/PROGRESS QUALIFICATION	153
PRODUCT SUPERVISION	227
ADMINISTRATIVE AUDIT	10
MARKETING	207
TOTAL	871

TOTAL COST OF QUALITY
EQUALS
COST OF NON CONFORMANCE
PLUS
COST OF CONFORMANCE
EQUALS
£4048K + £871K = £4919K
EQUALS
24% OF 1987 SALES VALUE

Figure 4.6 Example of initial cost of quality study results.

4.3.3 Employees

The third leg, and the one often most weakly addressed, is the employee leg. This is a sad reflection of our **Scientific Management** upbringing, since it must be obvious to anyone that we rely entirely on the support and active enthusiasm of our employees to deliver the goods in the other two legs. There is at the end of the day only one thing that differentiates one company from another – its people. Not the product, not service embellishments, not the process, not secret ingredients; ultimately any of these can be duplicated. There can only be one West Indies cricket team, Liverpool football team, Apollo 7 crew, etc. The Japanese have always recognized this and it is one of the reasons for their success in world markets – they place tremendous value on the integration of people with organizational objectives, equipment and processes. This is reflected in the much longer selection and induction processes employed by most Japanese employers and the careful way in which change is managed. Contrast this essentially prevention based approach with the traditional **fix-it** approach of most Western organizations, relying on probation periods and appraisal to weed out the poor selections and you start to get some feeling for why employees are often described as an under-utilized asset.

An interesting statistic we came across some time ago related to the distribution of professional jobs in Japan and the USA. Apparently, the USA has twice the population of Japan but they have half as many production engineers and 16 times more lawyers. What does this tell us about fixing versus prevention?

The employee leg forces us to address three key issues; communication, involvement and development. In fact the three issues can be used as a measure of an organization's maturity in the employment relationship. At one end of the scale, there is no communication in any direction except downward; involvement and development are unheard of ideas – employees hang their brains up in their lockers and collect them at the end of the shift. At the other end, companies have evolved communication into a fine art, empowerment is the norm and all employees have access to an open education system geared to continuous learning. The model is developed further in Chapter 6 as part of **P**articipation.

Many people confuse communication systems such as team briefings with involvement. Involvement, however is more than just the exchange of information. It is the gradual but radical delegation of control to those closest to the process itself. Self-managed teams, cell-based manufacture, autonomous work groups, high performance work systems, are all examples of true involvement. Quality Circles and problem-solving groups all play their part but the real issue is the level of responsibility for the job which is vested in those carrying it out. The change will require you to address some very difficult issues, such as the number of managerial layers, the level of trust, risk and failure you are prepared to contemplate and the shape of support functions such as purchasing, quality control and personnel.

Finally, none of this can happen unless you are prepared to invest substantially in the development of people to undertake the more participative and committed role this approach suggests. Training is only one aspect of development. Others include induction, linear movement, customer appreciation in its widest sense, secondment, participation on project teams, setting objectives and rewarding and recognizing progress.

Once again the state of play in the organization's performance relating to this implementation leg can be measured through the use of an employee attitude survey. IBM have long been users of regular attitude surveys to identify areas for improvement in the employment relationship and to track changes in employee satisfaction and feelings toward the business. This has often enabled them to nip developing problems in the bud. Chapter 6 discusses attitude survey design in more detail.

Similar principles to those outlined in the earlier section on customer perception surveys apply to the planning process for an attitude survey. Namely, that you should establish clearly your target population, the objectives of the survey and what you are prepared to do with the results. This final point is of even greater importance than when dealing with customers since sensitivity to the intervention is much higher among employees. Expectations of action will be raised. If you are not **seen** to act on the results, the survey can actually have a negative effect on attitudes. One major multinational acquaintance of ours got very low responses to questions on its performance in the communication field (we have yet to find the organization that achieves high scores on these parameters!) but failed to take more than cosmetic action. The result was that the following year's survey rated them even worse, despite the improvements they had made. Expectations had been raised and the result was that actual performance fell even further short of the new expectation level. This is illustrated in Figure 4.7. The employees' response is generally not based on comparisons but their own subjective view of the organization's performance. What they actually report is **the gap** between expectation and experience, which the survey itself had caused to increase.

This is not to say that we are not in favour of using attitude surveys – merely that users should recognize some of the hidden psychological impacts of their introduction. In planning the TQM implementation approach, an employee attitude survey will provide guidance for the key areas to be addressed in improving the employment relationship, but there are also other real benefits to be gained from the exercise. Most importantly, the survey demonstrates a concern for employees' opinions (with the caveat that some action must occur as a result of asking, otherwise entirely the opposite message will be communicated). The responses

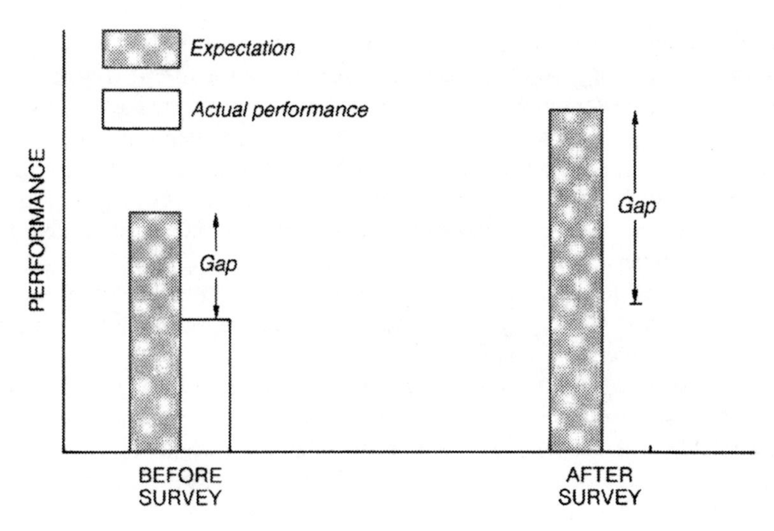

Figure 4.7 Customer satisfaction/expectation ratios.

provide management with indicators and justification for changes in policy or the introduction of new communication and consultation methods. They may also provide a range of suggestions for improvement in the case of **open** surveys which allow a certain degree of free responses. Underlying concerns in the workforce, which have not yet reached a stage where they surface through specific problems or grievances, can be spotted, as with the IBM survey usage.

In particular they can be used as before and after thermometers to gauge reactions to the introduction of new methods or the impact of organizational change. The effect of IBM's downsizing in the early 1990s on staff morale showed itself through the survey results and the company was able to monitor recovery after the programme through changes in the parameters in the survey instrument.

With the initial survey then, the focus is on assistance to senior management in planning the implementation strategy for TQM. Surveys should however be seen in the longer term as one of a number of instruments for the future participation of employees. Detailed design of surveys is therefore discussed in more detail under Participation in Chapter 6.

4.4 SUMMARY OF THE THREE LEG STOOL MODEL

The three legs of the stool summarize the principal focuses of attention in a TQM strategy. But the early steps you take and the degree of attention each leg receives will depend upon the reasons for change. If TQM is a response to customer or consumer pressure it should be that leg which is the start point – activities such as customer surveys are a useful way to show concern for the customer and identify the key changes to be made to customer service and relationships. If the driving force is waste, then a cost of quality study at the front end to highlight the need for attention to quality could be appropriate. If the force for change is dissatisfaction with working conditions and job content, then an employee attitude survey should again enable the direction of time and resources to the main problems. Use of these survey methods in parallel can give a very powerful indication of commitment, express a concern for the views of others and help the management team to direct the implementation process successfully.

Data collection gives the team charged with implementing TQM a base for rational decisions and creates early involvement in the process and some sense of ownership of the conclusions amongst those most affected. Customers and employees are in fact a free source of consultancy if we would just take the trouble to actually consult them. Price Waterhouse *et al.* are likely to tell you the same things at considerably greater cost. Figure 4.8 shows the main data sources available. Note the relationship with the three key implementation objects described earlier and illustrated in Figure 4.1.

4.5 CHOOSING THE RIGHT CLIMATE

We distinguished earlier between culture and climate. Climate can be affected by many things. Experience shows that the introduction of a TQM process needs to be carefully planned and timed. It is unwise and difficult

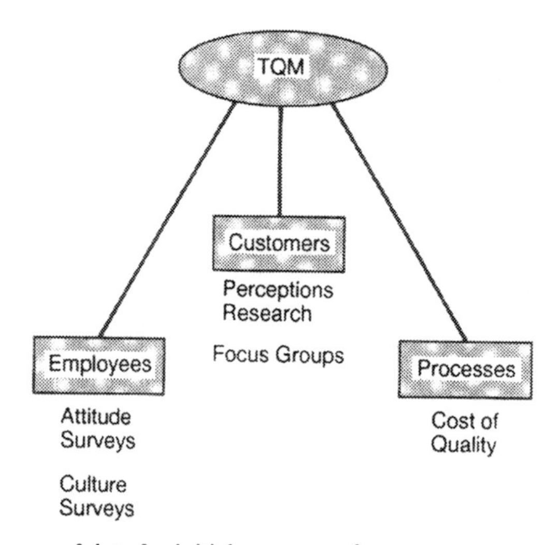

Figure 4.8 Sources of data for initial survey work.

to introduce these ideas during a redundancy programme, pay freeze, slump or other demotivating circumstances. On the other hand, this may be the time when TQM is most needed. The paradox is that the best time to introduce TQM is when you are doing well, but it also the most difficult time to persuade people of the need for change (if it isn't broken, why fix it?).

By the way, **time is not an issue**. There is never enough time to introduce TQM. People will always find reasons to put it off – too busy with current work or too busy chasing work – it makes no difference if your management team can truly perceive the benefits of Total Quality to themselves and the business. More about this in the next chapter.

4.6 DEALING WITH CULTURE

The two most frequently cited reasons for under-achievement in TQM programmes quoted in the 1989 survey by Develin and Partners referred to in Chapter 2 were **changing management behaviour** and **changing culture**. The first is in fact a manifestation of the second. Management behaviour is probably the predominant determinant of organization culture and it remains a fact that the nearer the top of the ladder you are, the more your behaviour influences the prevailing culture. Edgar Schein[4] said that culture can be sensed at three levels. Firstly the visible artifacts – physical conditions, reception areas, organization structure, office layouts and designs. At another level lie the stated values – statements made by senior figures, press reports, notices, annual reports and other publications purporting to represent the organization's attitude toward those with whom it interacts – customers, suppliers, employees, shareholders, the wider community. There is frequently an incongruity between the stated or **espoused** values and the evidence available to those who come to experience the organization's behaviour. For instance, most organizations' espoused values emphasize the importance of employees, yet day to day reality for many people lies far from the altruistic goals of the authors of organization value statements.

Precisely how out of touch are the espoused values with reality is a reflection of what Schein calls the **basic assumptions**. These constitute the third and most deeply rooted layer of organization culture. They are embedded in the organization's history, its traditions, its geography, the values and beliefs brought to it by its owners, managers and employees over many years. The older the organization, the more deeply embedded are these assumptions.

Perhaps the simplest way of thinking about these assumptions is as **the way we do things around here**. For most of us these are simply the vibes one gets when talking to people or watching how the employees of the organization behave. These vibes however are based on a number of very specific behaviours, systems and accepted modi operandi. At Danbury, we have developed a diagnostic tool to help organizations to get a feel for the main elements of **the way things are done** which measures employees' perceptions of key organizational and managerial behaviours. The results can give an indication of the areas of culture which are likely to be significant barriers to acceptance of the concepts of TQM. Strengths will also be indicated, allowing the management team to tailor its implementation strategy to take account of the existing culture and anticipate problems.

Studies have shown that the principles of TQM will only operate effectively in certain types of culture. Obvious examples are trying to institute careful process control approaches into dynamic businesses which traditionally fly by the seat of the pants or trying to get high levels of participation in an organization which is highly centralized or has an autocratic management style. The key culture changes implied by TQM are shown in Figure 4.9.

Encapsulating these cultural trends in some way is an important part of the **P**lanning process discussed later. Demonstrating the principles through behaviour and reinforcement is the job of the CEO and his or her management team. Simple and clear messages are required to communicate the meanings behind the more verbose and unwieldy value statements. Stew Leonard's two rules[5] are a good example:

Rule 1 – The customer is always right

Rule 2 – If the customer is wrong, see rule 1

Rule 2 is vitally important. The point it makes is that customers do not buy on the basis of what is true or right. They buy on the basis of what they **perceive** to be true. Thus it makes no difference whether the customer is right or wrong; if he thinks he is, then he is, because it is that which will determine whether he comes back.

For us the heart of TQM is the belief that:

Rules are made to be challenged

Promises are made to be kept

In other words, TQM is not just about achieving conformance of the product or service to its specification, it is about constantly challenging the status quo. James River Corporation of Virginia run a programme for employee and group innovations called **Find A Better Way**. This encapsulates the philosophy of continuous improvement and provides a motivation and reward system to create a self-perpetuating culture of improvement and success.

Readers will note that a substantial proportion of the cultural changes

CULTURAL CHANGE

From	*To*
Grapevine and secrecy	Open communications
Control of staff	Empowerment
Inspection & Firefighting	Prevention
Internal focus on rules	External focus on customer
Cost and schedule	Quality of conformance
Stability seeking	Continuous change and improvement
Adversarial relations	Co-operative relations
Allocating blame	Solving problems at their roots

Figure 4.9 Areas of cultural change implicit in a TQM strategy

implied in Figure 4.9 are related to management style and attitude toward employees. **Them and us** attitudes were not invented on the shop floor. They originated with management and are continually reinforced by the majority of managers in their day-to-day interactions. One foundry worker, who had been with a company that we were working with for over 15 years, told us that in the last 2 years he had been called by his CEO a lazy bastard, a shithead, a goon, a clown and a waster. We had asked people to say what was the most important thing they would change about the company if they had a magic wand. His answer had nothing to do with pay; it was a *bit of civility and an occasional pleasantry*. Even managers who think they treat everyone as equals frequently are not viewed as doing so. Underlying apparent equality is often an air of paternalism and condescension. One company took the bold step of including their shop steward on the top team TQM planning workshop. Applause all round. By the end of the evening however, during a lively debate, it was the MD poking the steward in the chest with his index finger. Had the physical contact been reversed, one suspects that the shop steward would have been out on his ear or told that he was out of order. This in-built view of people's motivation is, among all the cultural barriers, the one which most needs to be broken down. Konosuki Matsushita of the Electronics giant summed up our problem once when he told a group of Western businessmen;

> We are going to win and the Industrial West is going to lose out. There is nothing much you can do about it because the reasons for your failure are within yourselves.
>
> Your firms are built on the Taylor model (**Scientific** Management, piecework, simple assembly, routinisation, work study *et al*); even worse, so are your heads. With your bosses doing the thinking, while the workers wield the screwdrivers, you are convinced **deep down** that this is the right way to run a business.
>
> For you, the essence of management is getting the ideas out of the heads of the bosses and into the hands of labour. For us, the core of management is precisely the **art** of mobilising and pulling together the intellectual resources of all employees in the service of the firm.
>
> [Our highlighting – is management an art or a science? We think perhaps that it has for too long been treated as the latter].

The Industrial West is going to lose out? It need not be so. In the 1960s the car industry was studied by several behavioural psychologists[6] who mostly concluded that the motor industry was indeed different; that its employees were mainly pay-driven and that earlier concepts of motivation such as self-esteem and achievement needs simply did not apply. The experience of companies such as Rover Group with employee involvement and open learning initiatives has shattered this complacent view of people.

Trying to introduce TQM into an organization which has an incompatible culture is comparable to the heart surgeon carrying out a transplant. If the body and the implant are incompatible, the body will reject it. Gaining acceptance of the implant means both careful preparation of the body which is to receive it and cloaking the implant to make it appear as acceptable as possible to the body. Likewise in TQM, the body must be prepared by carefully outlining the need for change and the purpose of the process. The TQM implant must also be carefully shaped to take advantage of the organization's strengths and avoid too early a cultural clash which may lead to culture shock (rejection).

Changing the culture is partly the purpose of TQM itself, but it is also in many cases a necessary prerequisite to the attempt to install TQM. Preliminary management training in areas such as leadership skills and a thorough review of organization structure, management style and systems are often required. The prevailing culture can be identified through the diagnostic instrument mentioned earlier, but changing aspects of it can only be achieved through changing the causes of the culture. In other words, you cannot change the culture by addressing the outcomes, only by changing the system which causes the outcomes. In our workshops with senior managers and directors, we encourage them to carry out a diagnostic review of their organizations' structure, systems and style. Questions you should pose about the way your business operates which may give rise to the existing culture are included at the end of this chapter.

This issue of culture represents one of the fundamental differences between TQM and ISO9000 type approaches to quality. Cultural change takes place through a process similar to osmosis in plant cells. It is absorbed into the organization by exposing it to appropriate external stimuli, such as education, customer pressure, peer group and competitor behaviour and leadership by example. The word absorption is important; the dictionary definition is **the entire occupation of mind**. The word absorb itself is defined as **to take up and transform**. Unlike ISO9000 then, it cannot be implemented in the sense of any one set of specific actions, but occurs gradually over time as a result of a whole range of influences. In other words, whilst ISO9000 is installed, TQM must be absorbed. One way of visualizing this difference is depicted in Figure 4.10. Consider organizations as consisting of a number of layers, with the organizational or systems layers at the top and, lying deeper, the cultural layers. If you picture TQM and ISO9000 as two drilling rigs, to pinch the words of an old TV ad, TQM reaches the parts which QA cannot reach.

4.7 ELEMENTS OF SUCCESSFUL TQM IMPLEMENTATION PROCESSES

From our own experience and research, supported by available published surveys, literature and experiences of others in the consultancy field we have identified a number of factors which seem to be commonly present among companies who have been more successful than others in getting over this barrier of culture change. These are outlined below:

Senior management is totally committed and involved with the person at the top being seen to be the most visible and obsessed with quality in every aspect of the organization's performance.

The organization has a clear mission and a set of written or unwritten values which are constantly reinforced and referred to by management.

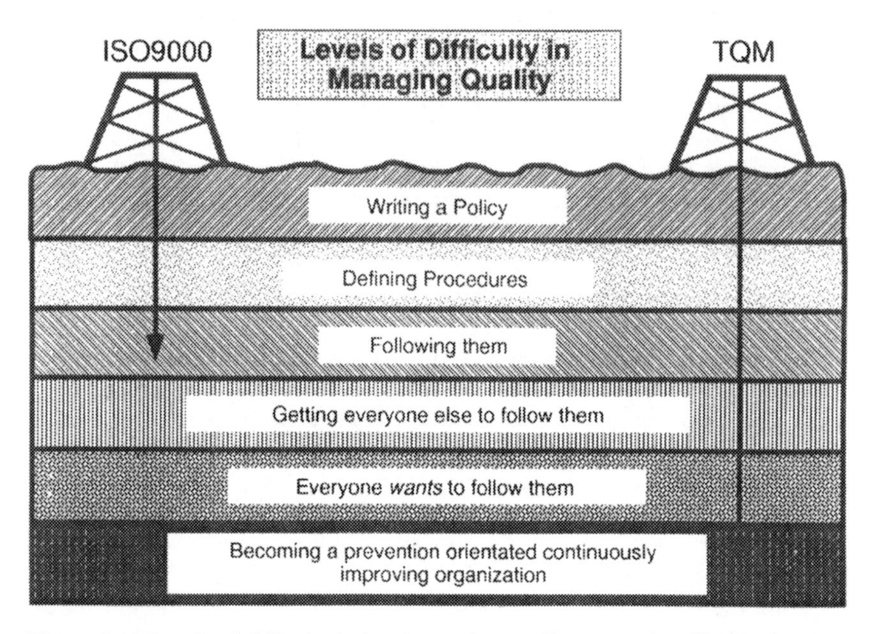

Figure 4.10 Levels of difficulty in implementing quality strategies. Deeper layers represent embedded culture.

The perspective of the customer, rather than the organization, is chosen at all times for driving action, although cost cutting may have been the initial driving force.

The internal customer/supplier concept is widely understood and applied through tools such as **Departmental Purpose Analysis**, which we will discuss later.

The company has a fundamental set of principles underlying the process, whether this be Crosby's four absolutes, Deming's 14 points, Juran's 10 steps or a home grown set such as we encourage our own clients to develop **and own**.

A previous attention to Quality Assurance through ISO9000 or other standards has helped to encourage cultural acceptability.

There is an organizational framework involving Steering Committees, Quality Improvement Teams, Quality Circles, Corrective Action Teams, Quality Councils, with a variety of roles and pet names. This framework is generally overlaid on the existing structure rather than adding to staff functions, although a full time co-ordinator, facilitator and/or trainer is often employed. In the longer-term successes, these structures have been absorbed into normal day to day management activity. Eventually, when the culture is embedded, the need for such props diminishes.

Activity is geared toward prevention rather than inspection and the true meaning of prevention is properly understood.

Corrective action is a formal system, often prompted into being by the relevant requirement under ISO9000.

All employees are trained in the use of measurement and problem solving tools and techniques. Critically, they are given the time, support and resources to use them.

High levels of employee involvement, radical delegation and flat management structures are in place.

The small team is seen as the powerhouse of the organization.

Training and employee communication and reward systems are given a high profile.

Middle management, those with theoretically most to lose (power, status and possibly position in the longer term), are brought into the planning process early.

The preliminary activities of assessing the current position and identifying objective reasons for change provide the information and understanding on which the planning process, explained in the next chapter, can be carried forward.

CHECKLIST/IDEAS MENU FOR CHAPTER 4

Identifying the need

Customer survey

Employee attitude survey

Cost of Quality survey

Organization culture check

Preliminary management training – needs analysis

Inter-department quality audit

Diagnostic review

Organization structure

- Do you keep discrete business units as small as possible?
- Do you have decentralized or centralized control?
- Are staff (non-productive) sections kept lean?
- Are responsibilities for support services as close to the user as possible?
- Are people encouraged to move between staff and line functions?
- How many layers of management between operator and CEO?

Systems

- Do recruitment and selection processes make use of modern aids to identifying competencies, skills and knowledge requirements, rather than on feelgood factors or the fact that the promotee was good at his last job?
- Job and person descriptions (simple, flexible, non-bureaucratic ones!) in use?
- Facilities for employee counselling?
- Performance reviews and **developmental** appraisal systems in use?
- Does appraisal encourage growth or create fear and distrust?
- Regular planned training and development for managers?
- Training plans for all employees?
- Customer care training for all front line employees?
- Communication systems –
- downward? (e.g. team briefing, newsletters, house journal, videos, telephone hot-lines)
- upward? (e.g. team meetings, suggestion schemes, problem reporting systems)
- laterally? (e.g. cross-functional teams, focus groups, lunch discussions, ad hoc 'huddles', direct contact between operators without need to go up the managerial tree)
- Regular surveys of employees and customers for their views?
- Are the business mission and values simply and clearly communicated to, and understood by, everyone?
- Employee share or profit participation schemes in place?
- Varied reward systems – cafeteria style without status symbols?
- Pay related to **appropriate** quality related performance criteria?

Style

- Customers' needs of prime concern in daily discussions about the job or problem in hand?
- Internal customer concept understood and reinforced?

- Internal customers treated with same importance as external?
- Promises kept routinely? (Add up the number of meeting minutes where a manager has not done what he or she said they would do at the previous one – you might be shocked. If your operators performed to that degree of conformance, what would you do? Another creator of **them and us** cultures.)
- Employees encouraged to **own** their work, quality and place in the success of the business?
- Scope for all employees to feel winners rather than a select few?
- Participation and consultation over decisions encouraged?
- Problems are managed by discussing elimination of causes rather than looking for someone to blame (change management versus blame management)?
- New ideas are encouraged and supported and well intended failure rewarded?

REFERENCES

[1] Reported in an article by Sandra Vandermerwe and Douglas Gilbert in *Business Horizons*; Nov/Dec 1989

[2] From 'Marketing Success Through Differentiation – of anything' by Theodore Levitt (Jan/Feb 1990). Copyright 1980 by the President and Fellows of Harvard College; all rights reserved. Quoted in Peters, T (1987) *Thriving on Chaos*, Pan Books in association with Macmillan, London

[3] Crosby, PB (1984) *Quality without Tears*, McGraw-Hill, New York

[4] Schein, Edgar H (1980) *Organizational Psychology*, third edition, Prentice-Hall, Englewood Cliffs, New Jersey

[5] Peters, T (1987) *Thriving on Chaos*, Pan Books in association with Macmillan, London

[6] Notably Blauner, R (1964) *Alienation and Freedom*, The University of Chicago Press, Chicago

5 Getting started

Positive commitment
Planning

5.1 POSITIVE COMMITMENT

All experts on TQM implementation talk about the need for top management commitment. In Chapter 3 we explained that change can be initiated at various levels in the organization. However, because of the degree of control (monopoly might be an appropriate term) which top management have over the policies and systems which enable or block attention to quality, the eventual conviction of top management in the value of the approach is essential. Obviously it is better if that conviction occurs naturally.

We use the word positive for two reasons. Firstly, because commitment doesn't begin with a P and it would mess up our Seven Ps if we used it! More importantly it conveys the fact that simply saying you are committed to quality is not enough. It must be demonstrated in actions as well as words, policies and mission statements. We have yet to meet the CEO who is not committed to quality, but the one who clearly and unequivocally lives by it is a somewhat rarer animal. Practical would have been an equally acceptable word.

5.1.1 Setting the cultural tone

Examples of this practical or positive commitment abound, however. The Managing Director of BP Chemicals led his company's TQM effort in a very special way. Trained by the Crosby organization in the UK, he not only attended the standard courses for CEOs and senior management, but also put himself through a further 5 days of classroom training to qualify as a course instructor. He then personally ran the in-house courses for his own management team; a role normally assigned to the Quality Manager, Quality Co-ordinator or Personnel/Training Manager. He consistently visited most of the further training programmes for lower management layers and contributed to the discussions about the aim of the programme. It is no coincidence that BP Chemicals had one of the most successful applications of the Crosby approach in the UK.

The Operations Director of Girobank personally opened all 73 of the company's quality training seminars. Senior managers joined in for periods during the day.

Colin Marshall, Chief Executive of British Airways attended nearly 300 of the first 430 **customer first** quality seminars personally to address the staff and answer questions. To emphasize his concern for quality he used to ignore or refuse any request from his managers which was not about customer service. His staff soon learned that if they wanted him to read something, they would have to build in a customer service angle and show how the customer would benefit.

Roger Milliken and Thomas J Malone of Milliken, the US garment manufacturer, reckoned to devote more than half of their time to the quality process in its early years.

Positive commitment is about top management leadership of the process. It is critical. Whilst we have observed many organizations where the drive for TQM has originated among middle management or other members of the management team, unless they have eventually been able to capture the imagination and support of the CEO, the original enthusiasm of those who wished to see change has ultimately waned. They have either accepted the situation or, more often than not, left the organization.

Among the CEO armoury for killing off the quality programme there are a variety of what we call non-quality snowballs. The further up the hill they roll out from, the bigger their impact on the people in the organization below them and the end customer. Some commonly used snowballs:

We really ought to have a meeting about this quality thing I suppose

What's the presentation about? Oh, quality, I've already been on one of those

Quality is all very well, but we've still got work to do

Do you realize how long it would take us to recover the man hours if we had all our people in team meetings solving problems?

Do we have to call it TQM?

I believe we have always done quality work

Can we make sure this (quality) meeting is over by 10 o'clock – I've got another appointment

I know we said we would meet weekly to discuss quality, but I just don't have the time

That will have to do; let it go

Don't worry, we can always fix it later if it goes wrong

To err is human, no use crying over spilt milk, mistakes will happen, etc.

These little asides are usually not even noticed by their utterers, but they have a major effect on the credibility of the process and on the staff's view of the commitment of top management. However, words are but one means of killing off the process before you have started. Even if you avoid the clangers and say all the right things you can still dislodge a snowball accidentally. As Tom Furtado, formerly of Pratt and Whitney, once said **employees watch your feet, not your mouth**. Support for the programme through the kinds of practical actions outlined earlier are important. Other rules for the CEO should be:

Never walk past a problem without doing something about it

Constantly challenge assumptions about organizational constraints

Keep promises routinely and religiously and don't make them if you can't. In our experience of guiding organizations toward implementation of TQM, there have been too many sad examples of CEOs who promised to attend or speak at the initial quality training sessions for their staff, who in the end found every excuse under the sun not to do so or to send a substitute. People are not fools. Even if those at shop floor level don't get the message immediately, the canteen and locker room talks with managers and supervisors will soon complete the grapevine. It is impossible to hide a lukewarm commitment to quality, but a real and consistent commitment from senior managers can be more infectious than the flu.

The job of a leader in an organization is to lead. This obvious truism is frequently forgotten; many CEOs think their job is to control and administer. Far from it; there are plenty of people capable of managing these tasks and they are probably already doing it successfully. The CEO's job is to make them feel uncomfortable, not assist in maintaining the equilibrium. Pushing back the barriers of bureaucracy, fighting the customer's corner in all matters, defending the employees, all require repetitive thrusts from the top. Consistency of leadership is paramount. In increasing awareness of the need for change, the chief executive needs to be constantly repeating the same message, orally and practically, whilst encouraging others to question the status quo. Finding a favourite phrase is one way of projecting the message. **Rules are made to be challenged, promises are made to be kept**, mentioned in the previous chapter, for example. **Under-promise, over-deliver** is another good one, from the Tom Peters camp.

At Anglian Water Engineering and Business Systems Limited, they invented a hypothetical cartoon character called General De Hassle (a play on de Gaulle) to front up their messages on quality, making the point that one of the main pay offs from the programme was the improvement in the quality of work life for employees. **De-hassle it** became a favourite expression.

5.1.2 Establishing a vision

The informal acts outlined above are, in our view, far more important, and hence far more difficult, than the formal acts of commitment required by the CEO and his top team. Nevertheless some formal actions to highlight the change of direction for the organization are necessary to provide a backcloth to the programme. These formal actions are in three areas: setting a quality policy, developing a business mission and determining the organization's value set or philosophy. These can then become formalized documents which can be communicated and referred to in everyday management actions. Many organizations have already developed such documents, perhaps as part of a marketing or other business direction setting programme. That should not detract from the fact that the roots of a TQM process must lie in corporate strategy. TQM cannot be divorced from corporate strategy; it is essentially part of that strategy and ought to be the vehicle for its implementation. The starting point for any TQM process must be the boardroom.

In one sense the quality policy can be seen as a special kind of value. The three statements together express the direction of the business; the mission says what you want to do and where you want to be in the market; the quality policy and values describe the way in which that mission is to be achieved. If the mission is the destination, the values and quality policy are the boundaries of the path over which you must not cross. It is possible to become market leader (a mission) by screwing your customers, making your employees' lives miserable and not giving a fig for the environment in which you operate. An article in a local seaside newspaper quoted a pier owner, under pressure from local residents about noise from a new helicopter pleasure facility he had introduced. 'The town is full of geriatrics and psychiatrics.' This man's value set is not hard to see. He may make the pier successful and himself rich (his mission) in the short term, but one questions the long-term effects of such actions.

We will deal with each of these three key statements in turn.

5.1.3 The quality policy

This serves two purposes. Firstly it indicates a change of direction by the organization in its positioning of quality in relation to the other two commonly accepted competitors for attention; time and cost. One can argue that time and cost are just aspects of quality. If we regard quality as conformance to requirements, then all activities can be said to have a time and a cost requirement. However, the practical reality is that these three concerns do compete for primacy and in the past quality has all too often lost out to the need to keep within budget or get the goods out on time. Management snowballs such as **let it go, it will do, we haven't time to put it right, never mind**, become part of the organization's philosophy by default unless challenged by some superior authority. The quality policy is that superior authority. It is a formal commitment which should be constantly referenced whenever this traditional fight for supremacy arises.

Secondly, the quality policy serves as a sounding board for the organization's examination of its structures, systems and management style. The diagnostic review referred to earlier should have highlighted the key barriers and change agents in the implementation process for TQM. Proposed changes should be measured against the Quality Policy and the question asked: **will it (the change) support or detract from our quality policy**? For example, the introduction of a new bonus scheme might be propounded. If this is volume related, it might encourage the quality policy to be sidestepped in favour of time. Profit sharing might encourage short term thinking and abandonment of quality in favour of cost unless the true relationship of quality to profit is understood by everyone.

The form which quality policies take appears to be remarkably similar, from the variety which have been published. Some organizations mix the quality policy in with the mission and values, but this tends to produce a rather unwieldy document which is not easily translatable. As with all corporate messages, the quality policy should be simple and to the point. Some of the better ones are shown in Figure 5.1.

The quality policy should be communicated to everyone concerned; employees, customers, suppliers, shareholders, etc. Many organizations incorporate the statement as part of their corporate image. One company even changed its livery to Total Quality Transport. Less ambitious actions might include putting the policy in sales and purchase documentation, having a plaque made, adding it to the employee handbook and putting it in the financial accounts.

Communication is the key. As with the other two statements of mission and values, the quality policy is useless, and possibly counter-productive, unless it is well communicated and used as a constant source of guidance by management. We have seen too many organizations which have issued such statements to staff, or stuck them on the wall, with no explanation, as though they were conducting a fly posting campaign for an election.

EXAMPLES OF QUALITY POLICIES

Rhone Poulenc Fibres Quality Policy

"It is our wish to deliver products and services exactly conforming to the needs of our customers. This means that each person in his duties, commercial, productive, technical and administration, is organised to eliminate non-conformities and do his work right the first time round".

Visqueen Quality Policy

"Visqueen will deliver defect-free products and services that conform every time to the requirements agreed with our customers.
We will ensure that our functions and processes are directed to enable all employees to achieve defect-free performance".

Omitec Electro-Optics Quality Policy

"It is the policy of Omitec Electro Optics Ltd to achieve TOTAL QUALITY performance in meeting the requirements of *external* and *internal* customers.
TOTAL QUALITY performance means understanding who the customer is, what the requirements are and meeting those requirements without error, on time, every time".

THINK: PEOPLE – CUSTOMERS – PROCESSES
'Partners in Technology'

Philip Crosby Associates

We will perform defect-free work for our clients and associates. We will fully understand the requirements for our jobs and the systems that support us. We will conform to those requirements at all times.

James River Design Products

We will provide products and services which meet the agreed requirements of our internal and external customers, first time, on time, every time.

Ciba Corning Diagnostics Limited

It is the policy of Ciba Corning to achieve TOTAL QUALITY performance in meeting the requirements of internal and external customers.
TOTAL QUALITY performance means understanding who the customer is, what the requirements are and meeting those requirements, without error, on time, every time.

Figure 5.1 Examples of company quality policies.

Mission

Most businesses are complex organizations, characterized by different interest groups with often conflicting priorities. Traditionally staff and line functions clash, as do sales with design, finance with personnel and production with development. Each of these groups have their own unwritten missions. For production it may be winning business with maximum run lengths, for sales it may be expanding into new markets, at the expense of simplicity of operation. For personnel it may be keeping a motivated and happy workforce, for finance it may be keeping wage costs down.

Often, the company has grown from being a small supplier in a single market to being a major employer with a diverse range of products and markets. The internal communication systems however may not have developed at the same pace, nor the ability of the management team to lead rather than administer the business. New people may have come in with no real knowledge of the business' origins or central product or specialism. Their goals may be quite different to those of the founders of the business. Other traditional conflicts of interest are between shareholders and directors and employees and managers.

This melting pot of objectives makes corporate goal setting difficult. The arrows are all pointing in different directions. No army could succeed if it simply relied on shooting at random and hoping to hit a target, let alone leaving the decision as to what the target is down to individual commanders in the field. Nevertheless, that is what many organizations are doing. Less than 10% of businesses in the UK have a business plan and even fewer a clear mission as part of that business plan. Yet the very survival of a business depends on its ability to plan ahead and obtain assurance of business continuity through knowing where its market is and what is happening to it.

Most of us like to think of our organizations as a team. A team is often defined as a group of people with a common set of objectives. These objectives are often implicitly understood at departmental level, but how their objectives fit into the overall plan of the organization is frequently far less apparent.

The mission statement aims to fill this void and should address just two key issues; the business you are in and where you wish to be in your market. Needless to say, in order to write such a mission, a thorough understanding must have been gained of just what that market is and what it perceives as *quality*. Toyota used to have a mission which translated simply as 'Kill Ford!'. This was unambiguous. It left you in no doubt as to their business, nor where they wished to be. And it was simple and communicable. We sometimes despair at the gobbledygook some companies come up with as a mission. It is often the subject of heated internal debate as the management don't understand it; and if they can't, what chance have the employees got? Had they been at Toyota, 'Kill Ford' would probably have read

'Toyota Motor Corporation aim to be world leader in the design and manufacture of automobiles for the domestic and international consumer market, with an emphasis on added value and innovative design and engineering specifications. We aim to achieve this through . . . blah, blah, blah'. Which one do you think is most likely to be remembered and followed by staff?

Having said this, what differentiates the Toyota mission from the average is its total flexibility (or absence of, if you prefer) on strategy. Too many people see their business simply in terms of what they manufacture or deliver and not in terms of their special skill or competence. Their missions usually reflect this lack of depth of thought. Marketing people call it seeing the product from the producer's point of view rather than from the point of view of benefit to the customer. AMEC Engineering shifted its mission from one which talked of carrying out capital and maintenance projects – essentially a reactive, specification driven business, to one which spoke of providing **solutions** to industrial engineering problems – a proactive, interpretative and value-adding strategy. 3M defined their core competence as *adhesion* which enabled it to vastly expand its product range and market whilst still sticking (no pun intended) to what it knew best.

Some other examples of reasonably succinct missions are given below:

AMERICAN EXPRESS[1]

To be the best provider around the world of quality service – bar none

DOMINO PRINTING SCIENCES[2]

The objective of the Domino Group of companies is to become world leaders in the electronic overprinting industry

THE TQM RESEARCH AND DEVELOPMENT CENTRE

Our mission is to assist organizations in both the public and private sectors to be more successful through the adoption and practice of Total Quality Management. We seek to become and remain the preferred supplier of TQM consultancy, training and research services in the UK.

The purpose of the mission statement is to give a clear central purpose and direction for the organization, but it is also a prime tool for goal setting and decision making. To be effective it must not be allowed to be simply an academic exercise, nor a piece of wall furniture. It should be constantly referred to by senior executives and used in activities such as individual performance reviews, departmental budgeting, marketing and sales strategy and production planning. The question should be constantly posed; does this action take us further toward our mission?

Values

The purpose of the mission is to communicate the fundamental *raison d'être* of the organization and a vision of where it seeks to be in the market.

However, this only tells us **what** is to be done, not **how** it is to be achieved. The purpose of establishing a corporate philosophy or value set is to ensure that the **way** in which the mission is being pursued is compatible with the organization's fundamental beliefs, including its quality policy. For example, a salesman may have an objective of doubling his sales; he could do this by making false promises or concentrating on new business at the expense of after sales support. A production manager may have a choice as to what chemicals to use in a plant cleaning operation. His cheapest choice may cause more pollution than an expensive material. If the values of the business toward the environment are clear, so should be his decision. If however, there is no such stated belief, he is more likely to take the minimum cost route, reflecting the unstated values of the organization. The value statement should guide everyone on the behaviours and beliefs for which the company wishes to be recognized.

Peters and Waterman, in *In Search of Excellence*[3], describe the few basic values that seem to be common to successful companies:

1. A belief in being the best
2. A belief in the importance of the details of execution, the nuts and bolts of doing a job well
3. A belief in the importance of people as individuals
4. A belief in superior quality and service
5. A belief that most members of the organization should be innovators, and its corollary, the willingness to support failure
6. A belief in informality to enhance communication
7. Explicit belief in and recognition of the importance of economic growth and profits

Depending on your organization's market, traditions and strategic objectives, specific values should be defined for some or all of the following areas:

- Customers
- Employees
- Shareholders
- Suppliers
- The local community
- The environment
- Internal colleagues

- Other parts of the 'group'
- Other specific important relationships

As with the mission and CSFs, the values should be simple, communicable, believable and capable of being demonstrated through the actions and behaviour of the organization's leaders.

Promoting an error-friendly environment

Apologies for this bit of jargon, but it really is important in displaying the positive commitment of top management toward the TQM process. As we said earlier, TQM succeeds only if the culture of the organization supports it. Creating an environment in which experimentation, and thus occasional error, is encouraged is one aspect of this environment. The other is the generation of willingness to declare problems when they happen. The error-friendly environment separates the change managers from the blame managers. The traditional response to problems when things go wrong is to find someone to blame. When a train crash occurs because of a system which allows for human error in overriding warning lights as a matter of routine, do we question the system? No, we jail the train driver for negligence. When a shop floor worker breaks a piece of kit, do we question the design of the equipment, the physical environment, the training system, the method of use? Probably not. It is usually easier to 'kick ass'. This has to change. As Deming says, for the single worker the rest of the organization **is** the system. It is usually the system which requires examination. Only when this is understood, and management take steps to drive out fear (Deming's point 8) from the workplace will you be able to take advantage of the opportunities which mistakes provide to learn a better way of doing things. Until then, the natural reaction to punishment will continue: find a way to avoid being punished. Hide the problem, cover it up until others have had their hands on it so they (management) can't find out which one caused it, fix it, blame somebody else. These actions cost organizations millions.

Error-friendly then means treating errors as vital data on which to manage improvement, using the knowledge of those nearest the problem to root out the reasons and institute corrective action. Leadership from the top through action to change the system when problems occur is the clearest way possible of demonstrating that you are serious about change. Directors must give unequivocal direction to line managers and supervisors about their responsibilities to help people do a better job rather than control and criticize.

5.2 PLANNING

The acts of Positive commitment serve another purpose. They provide a source of information for planning and lay the cultural foundations on which to implement the plan.

Information from initial survey work into areas such as existing organization culture, customer perceptions of performance and employee attitudes provide the raw data on which actions can be planned and direction given to the implementation process. This work should enable the organization to identify the barriers to, and levers for, the changes which TQM will bring. For example, if the culture survey indicates a low rating on the openness of decision making, obtaining the participation and enthusiasm of employees in problem solving will be difficult. The team will therefore have to address the systems, management style and organization structure which cause the process of decision making to operate as it does.

The mission, quality policy and values should have been developed and some initial communication carried out with employees. The need for several mechanisms of communication to be employed cannot be emphasized enough. For most organizations embarking on TQM, this is a big change and probably the most important commitment it will ever make. Putting a plaque on the wall and sending out a news-sheet to staff is totally inadequate. Several personal briefings, through cascade methods, corporate videos, journals, training seminars and CEO addresses should emphasize and reinforce the message of the strategy so that these key policies are understood and believed.

The planning process must focus on the way in which the mission is to be implemented within the guiding principles of the quality policy and values. One way to do this is for the top team to identify the business' **Critical Success Factors** (CSFs). The CSFs indicate the key issues of strategic importance if the mission is to be achieved. From these can be drawn specific organizational, divisional, departmental and individual objectives. The mission and CSFs enable all employees at whatever level to confirm the relevance of the objectives within their control. They are effectively subsets of the mission statement. Some organizations incorporate them into the mission statement itself by using phrases such as 'We will achieve this by . . .'. The National Freight Corporation Mission (Figure 5.2) is an example of this approach.

One way of defining an organzation's (or department's) critical success factors is to ask yourself 'What do we need to do well to be recognized as a superior supplier?'

Another way of looking at the question is to ask what represents value or excellence from the customer's point of view. This may not be immediately obvious from inside the organization.

For example a manufacturer of dog vaccines defined one of its critical success factors as:

> Reducing the age at which the vaccine can be applied to the animal

. . . since the earlier the vaccine could be applied, the greater the competitive edge, as veterinary surgeons could vaccinate more dogs before

NFC BUSINESS MISSION

NFC will seek to become a company for all seasons.
It will achieve this by developing a broad based international, transport, distribution, travel and property group with a high reputation for service in all its activities.
It will retain its commitment to widespread employee control.
It will have a participative style associated with first class results-orientated employment packages.
It will seek increased employment opportunities and a real growth of dividends and share values for its shareholders.

Figure 5.2 NFC mission statement.

an infection could set in, thus saving more dogs' lives and increasing their future business. Marketing alone was not the answer.

A local authority department servicing a Council Committee defined one of its CSFs as:

> Presenting information concisely but comprehensively to enable elected members to reach effective and timely decisions

In the former (product) situation, the CSF was easily measured since it was contained in the technical spec. In the latter case the department tracked numbers of queries from members as a measure of performance. A CSF should always be measurable or capable of being converted to traceable manifestations of performance characteristics.

The information gathered from customer surveys, such as those described in Chapter 4, should be used to ensure that the customer perspective is maintained when defining the critical areas of performance. One client of ours, who carried out customer research **after** writing the CSFs, was surprised to find that the one area (technical product support) was missing from their original list of CSFs. The key is listening to the voice of the customer wherever it sounds – through your sales people, user groups, employees in direct contact, or set up customer focus groups if need be.

This research of the customer and the market is essential if the CSFs are to really make the difference between your organization and its competition. **The definition of the CSFs is a pivotal point in the implementation process for Total Quality** and must not be rushed – too many activities which follow stem directly from them and will be wasted effort if you do not get it right here.

The CSF need not always be in a 'deliverable' area to the customer. One project management company defined its main CSF as the ability to attract and retain qualified engineers in a tight labour market. This completely refocused their attention to human resource issues which had been neglected in the past.

The rules for defining CSFs are that:

There should be no more than about eight. Four to five is ideal. Where companies have identified more than ten CSFs they find it difficult to translate the factors into action.

Each should be **absolutely necessary** to achieving the mission.

Collectively they should be **capable** of achieving the mission. There should be no missing elements

They should be input-oriented. Making a profit or hitting financial ratios, such as return on turnover (ROT) or return on capital employed (ROCE), etc., are end results. Businesses are not successful because they make profits; they are profitable because they are successful at their chosen expertise.

Some critical success factors should not be included. These are the things an organization must do or possess merely to be allowed to compete. Quality is rapidly becoming one such factor. If you are a supplier to a company with a *Just in time* (JIT) manufacturing system, on-time delivery is a prerequisite. CSFs should be confined to those issues which will help you win the game once you qualify as a player.

Some general examples of critical success factors are shown below:-

Best in class product quality
Identification of market needs
Shorter lead times than our competitors, for new products
Reliable suppliers capable of 100% conformance to specification and on-time delivery
Highly developed management skills
Lowest delivered cost
Lowest in-use cost
International reputation
Ability to change manufacturing processes quickly
Personal contacts with customers

Full product range availability
Order to delivery lead times
Innovative design concepts
Product customization

5.2.1 Shadow organization

In the case study included in this book an implementation proposal is shown in which reference is made to a **dual-track** approach. This concept is essentially based on the experience that most organizations require tangible proof in the short term of the efficacy of TQM and some payback to fund the longer-term changes in organization culture. To address this difficulty, we encourage organizations to establish a **fast-track** in which the TQM principles and techniques will be applied to some quick but significant wins by cross-functional teams. This gives the organization some learning experience and visible evidence to present to the sceptics whilst enabling the **culture change** process to make way down its own slower track.

The next chapter will deal with the fast-track in more detail. The culture change track, however, must also be managed. Unlike improvement projects which can be tackled by employees at the sharp end, the culture change track will inevitably require changes in structure, systems and organizational behaviour which only those in control of the purse strings and policies can alter. Having said this, however, involving employees in the process of suggesting and testing the necessary changes should be encouraged. In order to plan and manage the culture change process, an organizational framework must be created.

Several different models exist for this framework; Crosby's Steering Committees and Quality Improvement Teams, built around his 14 steps of continuous improvement, Juran's project teams, Deming's quality circle structures are some of the more popular methods. It is impossible to say that any one approach is best. In our experience, Crosby's 14 steps or a similar allocation of roles to a top team of directors and senior managers provides a useful starting point for an organization, which enables it to develop its own unique approach as it expands its knowledge of the process. We do not adopt some of the razzmatazz of Crosby's *Zero Defects Day*, but instead encourage organizations to examine their key business processes which affect the TQM strategy.

From this analysis, responsibilities can be assigned among the members of the top team. This book sets out the activities involved in implementing a Total Quality process under ten headings often used by our clients. The headings are not fixed or absolute. They are drawn from the most common key processes needed to make implementation successful. The initial studies of organization culture, employee attitudes or customer perceptions may also be used to identify the critical tasks which will need to be addressed.

It is important to note that, unlike the Seven Ps, the items do not occur in any chronological order. They are processes which are both interrelated and inter-dependent. They must therefore be co-ordinated to ensure that the critical paths are identified.

The 10-process master plan

1. Management obsession
2. Guiding system
3. Quality costing and measurement
4. Recognition and reward
5. Training and education
6. Employee participation
7. Corrective action system
8. Customer care
9. Business process improvement
10. Communication and awareness

The allocation of responsibilities among the steering committee members enables action plans to be drawn up and the individual plans co-ordinated into a master plan for implementation. Note that many of the issues in the plan require attention to the management systems in the organization. The steering committee is **not** a vehicle for **getting employees involved** so that the directors can get back to their jobs. The principal responsibility for quality improvement lies with management, who own the system. Nor is it a forum for solving problems as such; rather it has an enabling role in making sure that the systems, training resources and support are in place to allow TQM to take hold. Steering committee members should be bridge-builders and barrier-bashers, not bureaucrats and fixers.

The steering committee requires some administrative roles to ensure its effectiveness:

A **secretary** – to ensure that agendas and minutes of meetings are produced and the action plans brought together.

A **chairperson** – the traditional role but with an emphasis on wide debate rather than too autocratic control on the meetings. The members of the team need to be weaned off the traditional department based reporting orientation toward a common role of quality improvement in cross-functional business processes. In other words the chair needs to encourage horizontal thinking rather than the more usual vertical or functional thinking of the boardroom.

A **facilitator** – not essential but a useful role for some teams, particularly where the teamwork concept is relatively new. The facilitator is often seconded full-time to co-ordinate the TQM process and generally is skilled in problem solving, team dynamics and liaison.

One of the functions at the steering committee should be to provide an outside view of the effectiveness of the committee's functioning. That is, to look at the **process** of meetings rather than the outcomes of them. Or, as one of the participants on our courses aptly put it, facilitators are meeting technicians, not task technicians (thanks to Bob Lube of Dobson and Crowther).

The facilitator role is also valuable at the next phase of the implementation process in educating the workforce in TQM principles and in the various tools and techniques of continuous improvement. Organizations generally train a number of facilitators whose roles develop from messenger to trainer to team helper/adviser as the process of quality improvement matures. The appointment of facilitators is an important step in the implementation plan. They are in many ways the link between good intentions and actions and should be chosen carefully. A set of criteria for a good facilitator might be:

- Knowledgeable of, and enthusiastic for, the principles of TQM and its methods of application
- Capable of thinking broadly – helicopter vision
- Respected by peer group and senior management
- Able to withstand stress and frustration in a liaison role
- Ability to coach and persuade
- Clarity of speech and ideas
- Sensitive to interpersonal processes

In addition, one member of the steering committee, preferably the secretary or facilitator, should be the appointed contact with the consultants, where they are used.

Terms of reference need to be established for the team at the outset and before specific responsibilities have been assigned. A typical set of terms is shown below:

Company X Steering Committee – Terms of Reference

Overall purpose

To promote, oversee and co-ordinate the planning and implementation of TQM in all areas of operation within the company.

To create an error-friendly environment supportive of quality improvement.

Membership

The members of the steering committee are appointed by the company's board of directors. Membership is comprised of a secretary, a chairperson, the TQM facilitator and a senior representative from each department.

Relations

The steering committee maintains the following relationships with other bodies in the company:

Board of directors
Fast-track project team chairpersons
Chairpersons of departmental quality improvement teams

The steering committee reports to and receives its direction from the board of directors.

The steering committee consults with and reviews recommendations from the Quality Advisory Council. (See next chapter for details of the responsibilities of this body.)

Responsibilities

1. Document the role and duties of the members, chairperson, secretary and facilitator.
2. Allocate responsibilities among the members of the team for key activities within the TQM process.
3. Establish a master plan for the implementation of the various elements of the TQM strategy from the individual action plans developed by members of the team.
4. Establish sub-committees, as necessary, to assist in achieving the more complex tasks.
5. Review the information obtained from the surveys of organization culture, customer satisfaction, quality costs and employee attitudes in order to determine priorities.
6. Ensure the consistency of the TQM strategy with the business mission and values of the Company.
7. Establish and implement a training plan for the continuing education of senior and middle management in quality improvement approaches and techniques.

8. Establish systems within the organization to communicate management's commitment to quality and the progress of TQM.
9. Institute systems for the costing, measurement, auditing and improvement of quality at all levels.
10. Review and approve the terms of reference and objectives of cross-functional and departmental quality improvement teams. Review and approve plans developed by these teams to ensure their consistency with the business objectives of the company.
11. Monitor the progress, successes and problems experienced and make recommendations to the Board in relation to the removal of barriers to, or the allocation of resources for, implementation of the master plan.
12. Liaise with the Quality Advisory Council on reactions to the implementation of TQM and suggestions for improvement.
13. Help create the right environment for quality improvement by insisting on quality and demonstrating the principles by personal example.

Meetings

Meetings will be held monthly to review plans and progress. Additional meetings will be held as necessary. Meetings will be conducted in accordance with an agenda published by the secretary.

Minutes will be taken by the secretary to document all decisions and actions and will serve as an historical record. Minutes will be distributed to each steering committee and board member and posted on notice boards.

The meeting process will be monitored by the facilitator who will advise on improvements to team dynamics and the way in which the meeting is conducted.

The key roles in outline

The individual roles of the steering committee are set out below. As a memory key we provide a symbol to each role to convey the overall philosophy required. Each member may take on up to three of the elements, depending on the size of the company and may have to co-opt other employees in sub-committees to tackle the more onerous tasks such as education, quality costing/measurement and corrective action.

Management obsession

This role is usually adopted by the managing director, chairman or senior person on the team. Leadership by example is the key responsibility of this process. The person undertaking this task is both the beacon for quality and the organization's conscience.

Typical activities will be speaking at seminars, meetings and training programmes on the principles and importance of quality to the organization. **Walking the job** is an essential attribute; all managers must show a constant and consistent attention to quality and this role must lead the way, persuade, cajole and speak up for quality wherever it is being compromised.

The holder of this role must also show long term strength and determination in the face of sceptics who do not believe management are serious about quality. Thickness of skin and an aptitude for repeating yourself endlessly about quality are desirable personal attributes. Taking charge of the deployment of the quality policy, a process described later in this chapter, will also aid in keeping the overall focus of the organization and its management on quality.

Examples of management obsession with quality abound in the leading edge companies. It starts with an unswerving belief in the product of the organization and a determination not to allow any departure from the agreed requirements for any reason. Milliken's President, Tom Malone states that quality takes up the first 4 hours of every executive meeting and would do so even if the building were on fire![4]

Make sure quality is first on the agenda of every meeting at every level of the operation – there is nothing to which it does not apply.

The symbol for this role is the beacon.

Guiding principles

The complexity and all-embracing nature of the TQM process inevitably means that people will find difficulty stringing the components of the system together in their minds as part of a coherent strategy. Deming's 14 points and Juran's ten steps are examples of guiding systems which act as a road map to which people may refer during the various stages of implementation.

There are many examples of successful quality improvement processes utilizing these systems but they may not be suitable for every organization. It is important early in any implementation process that a clear set of guiding principles be established. There is no reason why the recommendations of any of the gurus cannot be adopted without alteration but **you must know why you are adopting them and what they imply.**

For this reason we recommend a thorough study of the authorities on the subject of TQM and the careful preparation of a set of guiding principles tailored to your own organization's culture and future needs. These must then be publicized and employees educated in the implications for their personal roles and responsibilities. Defining your own set of guiding principles also helps to ensure that the education process works off a value system that has been thought through by the management team themselves rather than by simply parroting the phrases of the wise and holy or presenting an off the shelf training package.

For example, one of our clients developed the following set of ten principles:

1. Quality means fitness for purpose; this means understanding and meeting the needs of the customer, not 'it will do'.
2. Everyone has customers – the internal customer is the recipient of your work.
3. Quality is total – everyone within the company is responsible.
4. Do it right first time, on time; set goals to improve every time.
5. Keep work methods and procedures simple and reduce variability in performance.
6. Improve effective communication between, and co-operation among, departments, customers and suppliers.
7. Make quality visible within an error-friendly environment:
   - report it
   - measure it
   - cost it
8. Develop a team approach to problem-solving; eliminate the root cause.
9. Recognize and reward successes.
10. Develop people by creating a continuous learning environment.

Total Quality is a culture, not just a system.

The symbol for this role is the map.

Quality costing and measurement

Measurement is the engine room of the Total Quality Process. Without measurement, quality remains invisible, management remain unconvinced and employees remain unmotivated to do anything about intrinsic work problems. The person assigned this task will need to work closely with the members responsible for business process improvement and corrective action. The principal task is to create the tools and environment in which quality is measured. Visibility is the key. Water companies in the UK, for example, now pay customers £10 cash if they fail to keep an appointment. This immediately makes it apparent to all concerned that there is a cost associated with poor customer service. If the cash were actually taken from the directors' wallets one suspects that visibility would be elevated from the statistical to the real!

Tasks will include identification of key quality data, definition and production of suitable measurement and display tools such as charts and boards.

Early analysis of customer perceptions and employee attitudes may also provide useful stimuli for action. This role will play a key part in the fourth and fifth Ps – **P**rocess control and **P**roblem identification.

One rule is fundamental: **Do not measure anything you cannot or do not intend to do anything about**. Measurement has only one purpose: to target action for improvement. If no action is forthcoming people will view it as bureaucracy. And don't measure too much. Paralysis by analysis is a common symptom of TQM. Use tools such as the Pareto (80/20) principle to focus on the critical few concerns.

Specific measurements need to be chosen by the people who will use them. It is not up to the steering committee or the person responsible for measurement to define what will be measured at the departmental level – only to create the conditions in which measurement will take place. The tools for measurement (e.g. charts, forms, graphs, pens, training packages, spreadsheet software, etc.) should be available immediately behind the training of employees in problem solving techniques; co-ordination with the training plan is essential.

Where some decisions on specific measures **are** required from the steering committee is in the key performance criteria for the organization and in establishing benchmarks for best in class performance. These should provide the macro-level indicators of overall performance and should be geared toward those factors which have greatest impact on the end user, such as delivery times, rejects, service satisfaction, warranty, credit notes, etc. Caterpillar for example, at the beginning of their quality improvement strategy developed an eight-point plan[5] for quality improvement. This is shown in Figure 5.3.

Visibility is also assisted by the use of the principles of quality costing (explained in Chapter 7). It is advisable to conduct an initial quality cost study to ensure that quality improvement does not take place in a vacuum and that projects are properly targeted to bring the greatest benefits. To achieve the study effectively, the person assigned this role must bring together a team and design suitable methods of data capture to enable each function or department to identify and report their quality costs.

The quality cost system is a progressive one. It is impossible to uncover all the facts at the start of the process. Probably only 40% of an organization's quality costs are apparent from existing reports and documentation. The first part of the quality cost team's efforts will be devoted to identifying these and their other major invisible costs. As time goes on, there will be a gradual increase in the number of items reported and the degree of accuracy of the figures.

However, it must be remembered that the quality cost system is not intended to be highly accurate. It is a barometer showing change in cash

EIGHT POINT QUALITY PROGRAM – *Milestones*

1 **A quality standard setting process** based on customer needs and competitive levels which incorporates product quality requirements into functional, reliability and durability terms.

	'82	'83	'84	'85	'86	'87	'88	'89	'90	'91/92
Setting Standards	OGO/SEGO setting reliability standards	PRM 200 Hr. DRF Stds. set	Field follow standard reduced 50%	PRM 201-1000 hr targets set DRF lowered 20%	•Appor'ment developed •Response targets lowered	90% of targets set for Expansion Line Products	Reliability appor'ment installed in production PRM			Product & process targets represent high customer preference

2 **A product design, testing/evaluation and manufacturing readiness program** which results in meeting quality requirements at first production.

	'82	'83	'84	'85	'86	'87	'88	'89	'90	'91/92
New Product	Reliability Growth installed	New Product Strategy initiated	Quality Strategy reviews begun	NPI Checklist used at production readiness	Quality guide book for NPI published	Reliablity appointment used for new designs	80% products meet targets			All products meet targets throughout life cycle

3 **A manufacturing conformance program** to plan and achieve conformance to design requirements

	'82	'83	'84	'85	'86	'87	'88	'89	'90	'91/92
Manu-facturing	Qual. Indicator NOPS includes DLY loss scrap & rework	Piece part conformance at 50% (all characteristics)	Conformance measure changed to Defects/1000	Worldwide plant conformance audits	Processes evaluated at 50% capability	CPK introduced as process improvement indicator	Internal Quality Certification established			All processes >1.5 CPK major commodities certified

4 **A supplier quality assurance-certification program** to supply parts and components which conform to the design requirements.

	'82	'83	'84	'85	'86	'87	'88	'89	'90	'91/92
Supplier	Certification discipline reinforced	Quality Plan guide in use	Certification video & brochure	Quality Evaluation Profile	60% of volume Certified	AQI video & brochure for suppliers	80% Certified SPC/AQI required			<0.5% Rejections >1.5 CPK AQI manufacturability Consolidation JIT

5 **A product support service to the customer** from factory and dealer which maximizes the return on his investment in Caterpillar product.

	'82	'83	'84	'85	'86	'87	'88	'89	'90	'91/92
Product Support				Dealer service quality survey started	Measures of CAT support to dealers defined	Dealer service surveys growing	Surveys introduced at subsid-iaries			Dealers & customers acknowledge product support as superior

6 **A field intelligence and response system** which causes rapid and complete correction of product problems.

	'82	'83	'84	'85	'86	'87	'88	'89	'90	'91/92
Field Intelligence	SIMS providing field data	Field follow feedback strengthened	Product problem system reviewed	Operations Grp reviewing past due Impact problems	Problem response time improving	Dealer inquiry system in place	Appor'ment used to idenify problems of sub impact			Field intelligence & response cause high customer satisfaction

7 **An annual quality improvement program** using quality education to improve the knowledge, skills, and attitudes of all personnel which impacts all corporate processes

	'82	'83	'84	'85	'86	'87	'88	'89	'90	'91/92
Education	Quality Education begun	•Juran training •SPC Course	•QUEST newsletter •Juran met w/officers	•Diagnostic Tools Course •Officers met w/media on Quality	AQI Matrix •Taguchi •QFD	Quality Institute initiated	Keynote at IMPRO 88 by CAT officer			Process improvement rate exceeds competitors

8 **An accounting and evaluation system for quality improvement** to help management direct resources to their most effective use by quantifying the cost of product defects and process waste.

	'82	'83	'84	'85	'86	'87	'88	'89	'90	'91/92
Non-quality Costs			Product related cost elements identified	Costs estimated at $400 M/Yr		Corporate AQI team established				Costs reduced to <$80 M/Yr

Figure 5.3 Caterpillar's eight-point quality plan (reproduced from TQM Magazine with kind permission of IFS Publications Limited).

terms from the effects of the TQM process in general and from the effects of individual quality improvement projects in particular. As with all measurement, only cost in detail those things you can or intend to do something about.

Implementation of the quality cost system is usually pump-primed by a training session in quality costing delivered by the consultant to the team followed by the initial study of costs by function or department. The person responsible for quality costing must prepare a plan for the system to lock into the plans of the steering committee member responsible for the corrective action system, which itself may be a source of quality data.

The symbol for this role is the arrow.

Recognition and reward

This is often a neglected part of TQM implementation, with few organizations getting much further than treating it superficially with issues like badges and certificates instead of addressing the real factors which motivate employees like personal verbal and public recognition, the way they are led, examples set by management, the opportunity to raise dissatisfactions and the kind of tasks in which they are involved. This role should also review the performance management and pay systems and challenge those which ignore or in some cases even contradict the quality process, such as pure volume related bonus schemes, new sales targets, etc. Do you recognize the firefighters at the expense of the planners? Batman is always a good deal more visible and gets greater kudos (thus feeding his ego and encouraging more fires) than those who prevent the problems from happening in the first place.

For years we have paid our salesmen on the basis of new business secured. Have you ever tried to get hold of the salesman who has pestered the hell out of you for the last 6 months **after** you have signed the order? What impression are you left with of the company? Salesmen are not to blame. They are victims of the system which management own. If we want them to care for the customer we must reward them on client satisfaction as well as sales value.

Financial rewards certainly have their place but it is the psychological climate created which really determines how the quality process will be received by employees. One client related to me how his MD had been shocked by the hostility of his copy room manageress when he had invited her to sit on a quality improvement team. What he failed to appreciate was that she had made four requests for his help in solving a mail delivery problem in the last three weeks and had received no reply or acknowledgement of her efforts. Listening is one of the most powerful tools of recognition a manager can employ. Cloth ears in management are one of the most potent forces for frustration and cynicism.

Another example of management failing to understand the power of personal recognition occurred in a company which achieved the ISO9000 certificate after a young and ambitious quality manager had sweated blood to establish the relevant systems and procedures. When the certificate was received by the company chairman he stuck a yellow 3M label on it to be redirected to the quality manager. No word of congratulations, no presentation, no formality, just the redirection of a piece of incorrectly addressed mail. They never could understand why he left to go to a competitor after spending the whole of his previous career with the company.

The symbol for this role is the rose.

Training and education

This is probably the most intensive of all the roles in the early stages of implementation and will require support from the rest of the team. One of the most common mistakes made in implementing TQM is to allow the training programme to take the place of the actions and behavioural changes needed. The other roles on the steering committee require equal attention, especially since there are certain parts of the process, such as quality costing, corrective action, measurement tools and rewards which need to be in place as soon as employees have passed through various stages of training. Successful implementation is heavily influenced by the fluidity with which employees can pass from training in the concepts and tools of TQM into applying them in their own work situations.

The first training requirement will be TQM awareness training in the concepts and philosophy of the process. This will need to be followed up with specific training in process control and problem solving methods. As competence in the basic tools grows, the training may expand into more sophisticated tools of problem solving and planning, directed at specific staff groups, such as Taguchi methods and Quality Function Deployment. Other possible training needs may be general management skills, team dynamics, job/process knowledge, etc. depending on the existing skills in the organization.

Companies such as Ford, Rover and Lucas have seen the creation of an open access education system as a natural development of TQM. In such a system employees are encouraged and supported financially in raising their educational qualifications, even if there is no direct link to the job. The belief is that a better educated workforce will be a more committed and thoughtful resource irrespective of the specific direction of the learning.

The symbol for this role is the tree.

Employee participation

The person carrying responsibility for this process is charged with creating the shadow structure necessary for employees to become actively involved in quality improvement projects; establishing Quality

Circles or Departmental Quality Teams, setting up cross-functional improvement teams and a Quality Council for example.

In addition, participative mechanisms such as share schemes, consultative processes in the general running of the business, suggestion schemes, etc. should all be considered under this heading.

Tasks will include the definition of standard terms of reference for permanent and ad-hoc teams, reporting and feedback mechanisms, publicity and rewards for successes in close collaboration with the members responsible for recognition and communication. This role is of such importance that it represents a complete stage in the implementation process and is discussed in much greater detail in the next chapter.

The symbol for this role is the pyramid.

Corrective action system

This role involves the establishment of a system for reporting defects and problems in a formal manner and creating a mechanism for channelling such inputs to the most appropriate medium for their resolution. This may be a new project team, an individual, a department team, a quality circle or even reference back to the sender. The objective is to provide an open access system in which everyone in the organization feels able and willing to declare that a problem exists. The creation of an **error-friendly environment** is essential if concerns are not to be shrugged off or swept under the carpet. Management must steel itself to accept criticism and treat all issues seriously and without defensiveness.

In essence it is an extension of the suggestion scheme principle, except that the suggestor does not need to have a solution, only a problem. Placing of boxes and corrective action form dispensers around the site is one way of making the channels of communication as open as possible. In higher tech organizations, the use of electronic mail systems and databases, may provide a user-friendly alternative to form filling. Such systems are employed at Perkins Engines plant at Peterborough to great effect. In lower tech situations the use of a facilitator to help people who are allergic to forms and computers to make an input is often helpful in ensuring total participation.

Organizations which have achieved certification to the ISO9000 standard will be familiar with the use of a formal system of corrective action. However, this tends to be tied to the manufacturing part of the business and responds to such stimuli as customer complaints, rejects, internal and third-party audit reports. Most of these stimuli are expensive. The people who really know where the problems lie are the employees – and they will tell you for free. If a corrective action system is already in place it should be overhauled to ensure it meets the following criteria:

It must be open to anyone, in or outside the company, to access if there is a problem.

It must have a defined set of quality standards in areas such as response time.

It must employ a range of different responses to direct the corrective action to the most appropriate problem solving medium.

It must not rely on the opinion of just one person to decide where the cause of the problem lies.

It must cover all areas of the business including finance, personnel, marketing, etc.

It must not be blame oriented. Management must get used to asking **what caused the problem** rather than **who did it**.

Tasks will involve educating everyone in the use of the system and devising suitable forms and explanatory notes or diagrams to encourage usage of the system.

Since there may be an overload of the system in its first few months, a sub-committee of the steering committee will need to be set up to process corrective action requests and ensure that there is an effective feedback and acknowledgement/reward system in place.

Other tasks for this person will include auditing the Quality System and reviewing the understanding and simplicity of the mechanisms established to ensure that an unwieldy bureaucracy is not created. The rule is that problems should be solved as close to their source as possible. They should not have to loop the loop through the corrective action system to get noticed!

The symbol for this role is the spade.

Customer care

The term customer here includes both internal and external customers, but the principal focus of the role is the external customer. The job of this member is to create opportunities to identify and exceed customer expectations. Externally, creating awareness of the quality process amongst customers, defining activities which help to raise understanding of customer reactions and perceptions of the organization, service standards, special customer events, special perks such as parking spaces, visits, identified contact staff, creating interactive processes, etc., all form part of what should be a constantly improving process of carrying the business to the customer.

This might sound like a good role for the marketing department but all too often they are more concerned with projecting the image of the organization to the customer rather than developing a real and sustained interest in every level of the workforce in understanding and meeting customer needs and expectations. It is the production staff who make the product and the receptionists and front line staff who experience the problems of poor service. These people have their hands on the product or service and can often relate much more easily to the customer's perspective.

The development of the **internal customer** concept is a key element in this role. The voice of the customer must not only be heard it must be translated back through the whole of the organization and its processes. There is a close working relationship between this role and that of Business Process Improvement.

The use of techniques such as departmental purpose analysis, internal quality of service standards, department benchmarking, focus groups and similar mechanisms all facilitate understanding of the important effect of internal customer transactions on the end product, external customer satisfaction and internal costs. This has the important side-effect of decreasing inter-departmental tensions as well as improving communication and job enjoyment. The internal customer tasks also overlap significantly with the employee participation role and quality costing/measurement.

The symbol for this role is the web.

Business process improvement

At the heart of TQM is the idea that improvement comes from addressing the **processes** which deliver the defective or inadequate end results rather than concentrating on the results themselves. The person responsible for this role needs to address business processes at both the macro- and micro-organizational levels. At the macro-organizational level, the principal tool is that of Business Process Re-engineering in which the critical success factors defined earlier are used to identify the organization's key business processes. Some of these may be in need of more than just incremental change and the potential of new technology and information systems need to be explored.

At the micro-organizational level, the use of Departmental Purpose Analysis, a technique also described in Chapter 7, can encourage groups of employees to look at the processes for which they and their teams or departments are responsible. This role works closely with customer care as far as the internal customer concept is concerned and with quality costing and measurement and corrective action. Many of the teams set up in the early stages of implementation may use the techniques which this role should champion. Ultimately the role of this person is to change the thinking in the organization away from reacting to the results or symptoms of problems to preventing the causes which usually lie in the processes used.

The symbol for this role is the chain.

Communication and awareness
This role ties in closely with that of management obsession and establishing guiding principles. The purpose is to continuously stimulate the attention of all those working for or interacting with the organization regarding your commitment and progress on the never-ending journey of quality improvement.

The role often comes under strain around 2 years into the process when the initial training programmes are over and the early projects have been completed. It is at this point, when management's attention starts to turn to other issues and complacency sets in, that the person responsible for this role needs to deliver a sharp kick in the pants to the process. If the process is allowed to wind down, employees' belief that TQM is **just another management fad** will be proven a reality.

Design of publicity, promotional materials, events, newsletters, progress reports and learning opportunities for employees, customers and suppliers is the central function. As Tom Peters aptly puts it in *Thriving on Chaos*, the objective is to create an endless stream of *Hawthorne effects** to keep quality uppermost in people's minds.

In the early stages of implementation, this role ties in closely with participation, education and training, the aim being to create a **critical mass** of support for TQM among employees. Awareness among employees of the cost of quality, customer perceptions, competitor activity, growth potential and opportunity all help to generate a belief in the need for change which is at the heart of classical change theory. Paul Gleicher[6] provides us with a useful model here which is very much at home in a TQM implementation plan. In algebraic form it can be shown as:

$$C = (ABD) > X$$

In other words **change** (C) only happens when the combination of

A – the need for change,
B – the clarity of where you want to be and
D – the knowledge of how to get there

is greater in total than the X factors – the barriers to change or the costs of change to the individuals or groups concerned.

* The experiments in work performance at the Hawthorne works of Western Electric in Chicago in the 1930s when Elton Mayo and his team demonstrated the effect of any kind of attention and stimulus to the work process in increasing productivity. For example raising the lighting increased productivity in a study group but so did lowering it again. The common denominator was the attention paid to the employees in the group, not the specific changes in the environment.

For example, a person will only change their job if:

A. There is a need to change – e.g. redundancy, desire to improve prospects, bored with existing job, etc.

B. There is a vision of the desired future state – e.g. I want to be a train driver, I want more money, etc.

D. There is a knowledge of how to get there – e.g. qualifications, location, how to apply, where to look.

All these factors in combination must be stronger than the costs or change of *X* factors – e.g. moving house, kids out of school, travel costs, being in unfamiliar surroundings.

In introducing TQM all of these factors will have to be satisfied; the reason for the formula being shown as (*ABD*) is that if any of the factors equals zero, then the product in the equation is still zero. Without the need for change being present, no amount of training, encouragement and visions of the future will cause lasting change to occur. The TQM implementation plan should therefore ensure that each of the factors are addressed by the steering committee:

A. Creating appreciation of the need amongst those affected by the TQM initiative as discussed above.

B. An ability to see how the process will operate – the manager might show the staff another site or company where TQM has been installed or demonstrate the practical and cultural aspects of the methodology by personal example. The use of a **fast-track** project as suggested in the next chapter is another way of painting a picture of the future state. Some companies have introduced pilots, such as in moving from assembly lines to cell-based manufacturing systems.

D. The knowledge of how to get there – instruction and guidance by senior managers supported by thorough training in the principles, benefits and methods of quality improvement.

These factors must then outweigh the costs of change, many of which may be psychological. For example, fear of failure, apprehension over training, threat to traditional power bases (trade unions, supervisors), poor literacy, concern over job security, change itself, exposure of capabilities. Other costs may be more tangible, such as time, effort, energy, additional work in the short term, possible loss of overtime if fewer problems arise and there is better planning in general in the organization. In some organizations moves toward just-in-time and other production efficiency programmes as part of TQM have led directly to job losses. This will be the biggest *X* factor if it is present. As we observe in Chapter 6, there may be a need for a wholesale rethink of job security, redeployment and redundancy practices to cope with this factor.

In reviewing membership of the steering committee, this role should be re-assigned whenever the holder starts to run out of ideas.

The role should also be examining communication in a wider context, in collaboration with the member responsible for participation. Employee involvement will only succeed where information flows freely and openly. If the culture is to keep decision making behind closed doors and to carefully vet information, then this will have to change if TQM is to flourish.

The symbol for this role is the river (the more dams you have, the greater the damage when one of them bursts – see Chapter 6)

5.2.2 Developing a LIMP (local implementation master plan)

As can be seen, the processes outlined above are not chronological but overlapping and developmental in nature. Individual plans are best prepared with the involvement of a facilitator or consultant. Alternatively, the steering committee may help one another by brainstorming some ideas under each heading which the person responsible for that element of the plan can use as an ideas bank. Some sample forms for individual action plans are shown in Figures 5.4 and 5.5. The form in Figure 5.4 is used as the overall plan for the process concerned and consists of a series of tasks or actions to be accomplished. The form in Figure 5.5 develops the separate tasks into a more detailed plan for their achievement and highlights where help will be required from other members of the steering group.

One person, usually the facilitator or secretary, should be made responsible for collating the individual plans from the members of the steering group, and putting them together into an overall plan for the implementation of TQM in the organization. This is not as easy as it sounds. The difficulties usually faced by the person responsible for this task are the vagaries of different people's use of the English language, a poor understanding of their roles by some members of the steering committee, one or two individual members failing to give the initiative their full support and not submitting plans or at best just putting in a few woolly intentions.

TOTAL QUALITY IMPLEMENTATION PLANNING SHEET

Process:- CUSTOMER CARE

Primary Person Responsible (PPR):- R. F. JOHNSON

ACTION	RESOURCES/SUPPORT REQUIRED	COMPLETION DATE
– Customer care affects and applies to both staff as customers to one another and/or ultimate customer (client).		Mid May
– Prepare end of project critique.		
– Prepare internal project critique that will facilitate the review of suppliers meeting the needs of customers.		End April
– Prepare checklist that can be used on visits to the customer which when completed will facilitate feedback to management and project groups.		End April
– Establish "Departmental purpose assignment" as formalized technique to be used following training in "Pat Pack".		
– Prepare lost project de-brief form that will facilitate feedback to management following job loss.		Mid May
OTHER PROJECTS BEING CONSIDERED		
– Suggestions on Car Parking		
– Customer Opinion Survey		

Figure 5.4 Example of completed individual action plan for steering committee member. (Thanks to AMEC Design & Management Limited.)

TQM IMPLEMENTATION PROCESS PLAN

Process Element: **QUALITY MEASUREMENT**

Person Responsible: **R SANSBURY**

Purpose of Process:

* To make quality visible and to gather information to make quality improvements.

Objective No. 4

* To make visible the progress and results of the quality improvement process.
* To provide adequate instructions on how to collect, analyse and display quality information and improvements.

Actions:

* Draw up a TQM Guidance Instruction for the collection and analysis of key quality measurement information, also for the display of quality improvement data.
* Define the information required as a result of the analysis of key quality information.
* Design and print suitable forms for the output of the analysis of quality measurement information.
* Design and print appropriate displays of quality improvement information.
* Provide details of quality improvements to all relevant recipients.

Quality Requirements:

* Ensure the TQM Guidance Instruction is adequate for the collection and analysis of key quality measurement information.
* Distribute quality improvement information to all relevant parties.
* Adequate display of quality improvements in appropriate, prominent locations.

Links to other Steering Committee Members' Plans

Element	**Person Responsible**
Quality Cost	Derek Norris
Communication and Awareness	Bob Johnson
Recognition and Reward	Tom Stringer

Figure 5.5 Individual action plan in detail. (Acknowledgements to AMEC Design & Management Ltd.)

For the plan to be effective it must move through three phases from:

1. **Good intentions** – stated commitments, lists of ideas and tasks to be accomplished. To:-
2. **Plans** – tasks allocated to specific members of the team. To:-
3. **Promises** – timescales agreed by when the plans will be accomplished.

The best way to accomplish the task is to set a date for receipt of the individual plans by the facilitator. The chairman of the group, the CEO and the individual charged with management obsession, should all aid the facilitator by urging their colleagues to complete the plans and making a point of talking over progress with them during the intervening period. Setting up some informal two or three person joint planning sessions among steering group members can help to ensure no-one can walk away from their responsibilities, and can provide support for members of the team who are unsure of their role.

When the plans are received, the facilitator should highlight any duplications, overlaps and areas requiring clarification and produce a debugging list as an agenda for the next meeting. Some tasks may require the establishment of small teams to manage them. Typically, the cost of quality study, the initial training programmes and the management of an **open** corrective action system are all time consuming and also benefit from the contribution of staff from all parts of the organization.

The Cost of Quality study needs input from all levels of the structure; the people who actually do the work know more about what goes wrong with it than any of their managers or supervisors. Setting up a team from all departments and within each department to contribute information to the Cost of Quality study spreads knowledge of the need for change as people start to see that much of their activity could be improved and frustration saved by the impact of TQM.

The training programme often presents a development opportunity for plateaued employees or rising stars in the organization to broaden their horizons and knowledge of the business and create skills for themselves in areas such as presentations, learning and managing change as well as creating a whole tier of ambassadors for the quality improvement process.

The corrective action system, like the cost of quality study, relies heavily on the willing participation of employees and the creation of an error-friendly environment. If these systems are developed behind closed doors, the information and ideas likely to emerge will be limited by fear of how they will be used. Unless this fear can be overcome the exercise will be only partially successful. This is one reason for setting up a sub-group. The second reason is that it is quite possible, at least initially, that there will be a significant influx of corrective action requests which will need more than one person's time to examine and redirect.

A mixed group of technical, managerial and operational staff can also

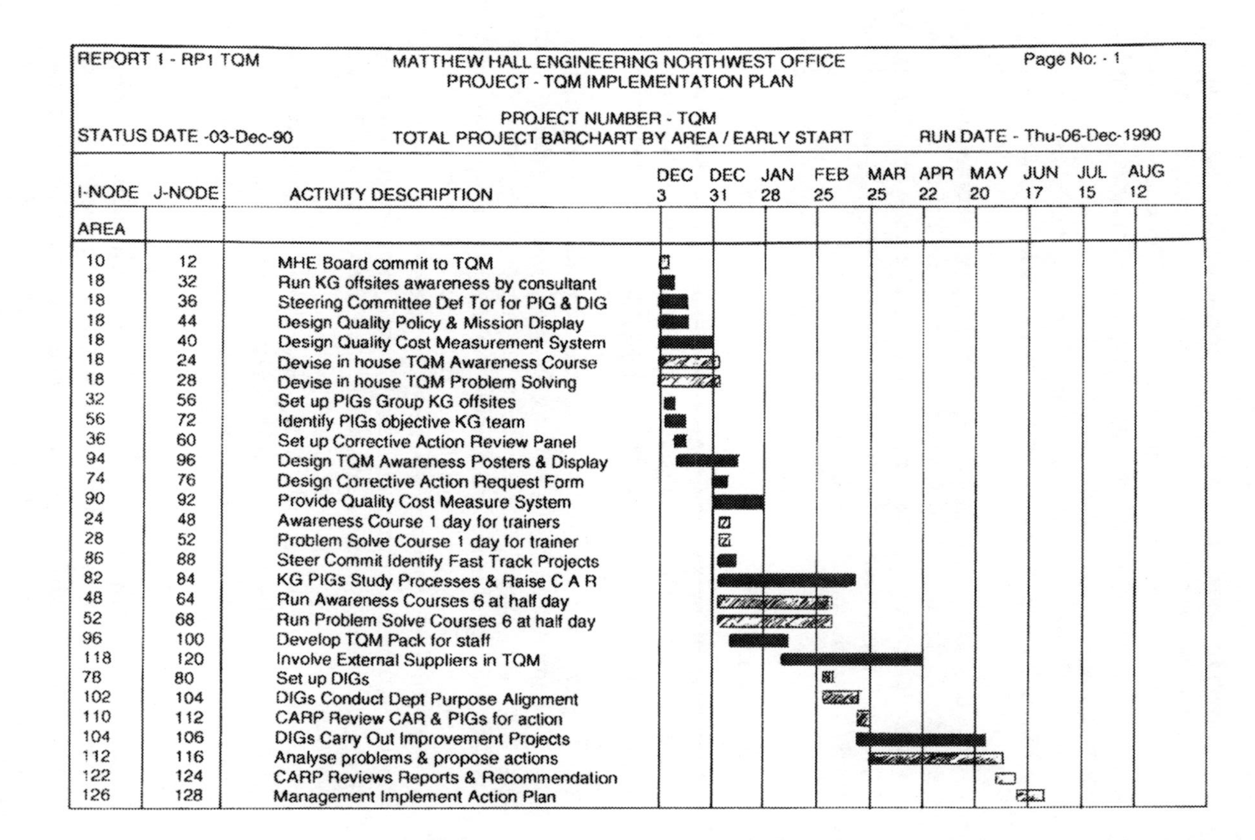

Figure 5.6 Example of a local implementation master plan. (Acknowledgements to AMEC Design & Management Ltd.)

help to overcome the traditional finger pointing which commonly goes on between functions and levels of the organization. Again it is developmental in nature as appreciation of wider processes are engendered together with an understanding of the nature of problems. This group must be trained in the principles of variation and techniques of problem solving but are not themselves problem solvers.

The final plan may be displayed as a Gantt chart or a matrix of actions and responsibilities. An example of an outline implementation plan is shown in Figure 5.6.

Once the LIMP is assembled, the steering group should audit it against the checklist at the end of Chapter 4 and the results of any surveys of organization culture, customer perceptions and employee attitudes. Have the key concerns been addressed? If there are any gaps, these should be filled in and the plan completed.

5.2.3 Changing the team

Membership of the steering committee should ideally start with the top team at the location to demonstrate management's commitment to the process (although there have been successful cases of delegation provided that enough constant interest is shown by the senior group).

However, once the process is up and running and the initial awareness and problem solving training is complete, the team's membership may be changed, in whole or in part or on a planned rotation basis. This in itself is a stimulation to the process and spreads employee involvement. That does not mean that top management's role is at an end; quite the contrary. As people who have been through it, they should become the champions, mentors and encouragers for those who follow and constantly question and stimulate activity among the new team members.

Even the roles themselves do not have to remain static. As explained at the beginning, the headings are suggestions only. Other process definitions may be more important or suitable for different organizations. Some other roles which might be assigned are listed below with details of their main responsibilities.

Supplier conformance: to review the external supplier base and the organization's policy and systems for supplier selection. To reduce the supplier base and establish single source partnering arrangements where appropriate. To institute a system for supplier assessment, rating and problem identification. To establish communication and problem elimination mechanisms with key suppliers. To review the opportunities to involve suppliers in product or service development and quality improvement initiatives.

Goal setting: a combination of the roles of measurement and process management. This role aims to establish a culture and methodology for continuously updating measurements and improvement targets at corporate, departmental and individual levels of the business.

Special events: a refinement or sub-set of the communication and awareness role, but taking responsibility for more detailed planning of special events such as open days, supplier quality conferences, customer visits, employee awareness presentations etc.

ISO9000 certification: many organizations see the installation of a Quality Management System to the British and International Standard as a prerequisite to TQM. However it can also be treated as an improvement activity integral to the TQM process itself. In such cases, the responsibility for obtaining ISO9000 certification may be allocated to a member of the steering committee. A similar rule could be developed for the Investors in People Standard.

5.2.4 Communication of the commitment

One of the reasons cited for the poor performance of some TQM programmes is lack of belief in management's commitment to the process. This stems mainly from a failure to communicate that commitment. Managers, in the UK in particular, get extraordinarily queasy about standing up and being seen to positively support something without reservation. Whether we do this out of insecurity or a fear of failure is not clear, but whenever you mention missions and values to a group of managers there is always a curling of the lips, either snarl or grin, from some of those present. Values are not seen as macho, uniess they are those you would not like the public or your employees to know about.

Somehow, we have to get comfortable about our beliefs and be prepared to live by the documents we put our names to, even if this means taking some flak initially from the sceptics. Mitch Kurzawa, at Chrysler said 'if you wink at the quality improvement process, your people will find you out and consder it just another programme for management'. He was right. Far too often, the organization's mission and values are referred to in just such a way, even by those who wrote them. Little wonder then that they are not believed by employees.

The James River Corporation of Virginia issue all managers with a **black book** containing the company mission and values together with detailed guidance to managers on their interpretation. However these are not simply disseminated but are the subject of induction and training programmes to instil the practical realities of the statements into their managers' behaviour. It is not unusual for the president and senior vice presidents of the corporation, when faced with a management problem to turn to the black book to gain guidance on a policy decision and to use it as a learning aid for line managers. It performs much the same use as a dictionary does for a typist.

Whilst the mission can be translated into action through the use of critical success factors as described earlier and the more traditional structures of performance appraisal and objective setting, the quality policy and values need to be cascaded down through the organization in other ways. Soft methods such as leadership behaviour, story-telling, myth creation play the major part in this but can be supplemented through a formal structured approach known as quality policy deployment (QPD). In essence, the process, which derives from the Japanese system of Hoshin kanri or policy control, is not dissimilar to the Management by Objectives approach, with the exceptions that the departmental objectives stem from the quality policy rather than the business objectives and the process is more team based than MbO which emphasizes personal objectives.

5.2.5 Quality Policy Deployment

The term Quality Policy Deployment is often confused with Quality Function Deployment (QFD). The latter is principally a design process which is discussed further under Permanence in Chapter 8. The direction of activity is also different, in that QPD follows a vertical route up and down the hierarchy, whereas QFD moves horizontally between functions to tie in customer needs to design, marketing, engineering, procurement and manufacturing concerns. The two are however complementary and together form the basis of many successful quality improvement strategies.

Quality Policy Deployment is a long-term strategy, typically being focused on a period of 8 to 12 years. It is a cascade process, translating key quality parameters, such as cost leadership, reliability, delivery speed, design cycle time, into departmental and team objectives throughout the organization.

It is supported by a macro-level application of Deming's (really Shewart's) continuous improvement cycle - Plan–Do–Check–Act. Although it is essentially a planning tool, it incorporates review procedures to constantly adjust and update the plan to reflect changed circumstances, barriers and reactions to the deployment of the policy further down the organizational tree. It is involvement oriented and therefore is conducted within a framework of participative meetings in which employees at all levels contribute to the policies and their achievement.

Unlike traditional MbO processes where objectives fall out of those above, without any consideration of whether those above have got it right, the flow in QPD often backs up several levels to review the correctness of earlier policy decisions in the light of input from those nearer to the customer, the product or the process itself. Have you ever been on one of those fun-house rides where you have to climb a two part moving staircase, moving in different phases? You stand on one side or the other; sometimes you are moving up and sometimes down, but eventually you get to the top, providing you make the right choices about when to move from one side to the other. QPD is like this moving staircase, taking you two steps forward and one back to check on the effectiveness of the policy as well as the efficiency with which it is implemented.

A typical Quality Policy Deployment, over a timescale of 1 to 3 months, may follow the route described below:

1. The Quality Policy is reviewed by the management team and the key parameters (e.g. service, reliability, price, delivery, etc.) are identified.
2. A 5 year or longer plan is drawn up for improvement in these key parameters (see Figure 5.3 for example)
3. A middle management conference is held to review the parameters and identify the main departmental activities which affect them.
4. Each department shadows the Corporate Improvement Plan with its own improvement objectives in those areas which critically affect the quality parameters. Recommendations and suggestions are made on the original quality parameters to reflect the departments' ability to achieve the standards set and their views on the relevance of the parameters to their own knowledge of customers' needs.
5. The top team reviews its plans in the light of feedback from its middle management and the revised plans are documented and passed on to the departmental heads.
6. A policy review between the functional heads and their staff is held to review the targets and establish short and mid-term priorities for improvement over 1 to 3 years. Recommendations are again made where the targets are considered unrealistic or irrelevant.
7. A series of team discussions is held between line management, supervision and employees to communicate the plans and add to the detail. Barriers to implementation are identified and plans made to cope with them.
8. A second policy review or debrief is held between the functional heads and those involved in step 7 to review the feedback and amend the short to mid term objectives in the light of the information and responses received. From this an improvement list is produced with responsibilities allocated to departments, managers, quality circles, project groups and other appropriate bodies to implement.

A half-yearly review at corporate level and a quarterly review at department level is held to assess progress against the short term plans and adjust the medium and long-term plans as necessary.

An annual review of the plans is also held by the top team to look at changes in marketplace needs, competitive information, customer survey data, audits and information from other feedback systems. The data are used to revise the long term strategy and recommence the cascade process of the Quality Policy.

The policy deployment and its progress is communicated to employees through a number of different media. Communication processes such as team briefings and quality newsletters may be used but these are supplemented by a departmental register of the policies and improvement targets, together with appropriate graphs and charts showing progress, which are updated regularly and open to inspection by employees and management alike.

Policy deployment through this method helps to ensure common objectives, which employees and managers have bought into, as well as identifying areas of process control in need of attention such as automatic defect prevention, inspection, elimination of variation in methods, materials and equipment in use, feedback, control points and self-checking needs for operators. Measurement charts on key quality indicators can be set up to support the areas where improvement potential has been identified.

The QA manager or TQM facilitator may be called upon to act as a co-ordinator in the process, ensuring meetings, reviews and communication stages are held as well as gathering and condensing progress reports for the steering committee and executive board. These reports should cover progress against agreed plans, variations from plan, constraints encountered (and any action required by senior management to remove them), successes achieved, responses from employees, customers and suppliers to related improvement initiatives.

Internal and third party audits or assessments of progress can be utilized effectively, similar to the kinds of mechanisms employed in ISO9000 procedures. This helps to remove the *halo effect* (the tendency of people to pump up their own departments' achievements). Internal audits also help create a competitive spirit among employee teams if handled sensitively by management and apply additional momentum to the process which can easily run out of steam unless directed and constantly pressurized by the top team.

Further indications of progress can be extracted from defect logs and improvement or change logs maintained by each department or team to note significant events related to the key improvement areas established in the local plans. These provide useful anecdotal evidence for problem solving activity as well as records of changes to the process or system which have been instituted as part of the improvement plan. Some organizations display before and after photographs to illustrate the results of improvements.

5.2.6 Planning against award criteria

An increasingly popular (albeit, we suspect, time-limited) method of planning and measuring progress on Total Quality is the use of national and international award programmes such as the Deming Prize, the Baldrige Award and the more recent European Foundation for Quality Management (EFQM) and British Quality Association (BQA) awards. The details of these were set out in Chapter 3.

This approach can be summarized as a set of advantages and disadvantages as follows:

Advantages	**Disadvantages**
• Creates a common set of criteria against which other organizations following similar routes can be benchmarked	• There is little opportunity for staff to buy in to these criteria except to be told they will follow them
• Puts in place a model for the organization to follow if it is struggling to define its own	• Can be a cop-out from applying original thought and conflict with organization culture
• Is based on accumulated best practice and research of sucesscessful businesses	• Best practice is a notoriously fluid concept; models, once written and infrastructures applied to control them, tend to resist change. Today's progressive model may be tomorrow's lead weight!
• Can lead to a major internationally recognized award eventually	• Only a few can achieve them
• The effort required to apply for the award acts as a stimulus for action and keeps management's eye on the ball	• The effort required in preparation and assessment takes considerable man-hours and cost and does not of itself directly benefit the customer or employees

5.2.7 Making the time

A common plea from people involved in the early stages of TQM is 'OK, all these things are great (QPD, QFD, Quality Circles, measurement, etc.) but we don't have the time to do everything we would like; we still have to do our jobs'. The simple answer is that nobody has time to do everything, but everybody has the time to do anything. Quality improvement is not compulsory (neither is survival, says Bill Deming). A manager's job is typified by never having enough time, people, money or whatever to do everything he or she wants. It is a matter of priorities. Sir John Harvey

Jones is fond of saying that the job of a manager is to make the best of what he or she has got. If you say we can't afford the time for our employees to meet to discuss quality, what you are really saying is we can't afford the time to get better; we will let our competition overtake us while we stand still.

All the initiatives discussed in this book take more of one resource than any other and it is not money; it is time. Top management must make clear right from the outset its expectations of itself and others in regard to the amount of time which will be devoted to the quality improvement effort. Make no mistake; if you are a member of the executive group and the steering committee, you should be thinking in terms of 20 to 40% of your available time being spent on activities related to the quality improvement process in the early days. Less than 20% is unlikely to create the belief that you are serious nor will it generate the critical mass of support needed for the process to become self-sustaining.

Equally important, line management has to know that it is OK for them to spend their time consulting and involving employees and working on plans which are not directly production related. In most cases, managers perceived role expectations are about maintenance activity; how much have you produced this week? If you are asking them to embark on a different set of objectives they have to feel comfortable that they are not going to be blamed for temporary production inefficiencies while they are undertaking some of the long term value-adding initiatives which are the trademark of TQM. In some industries, such as construction and design/ project management, it may be appropriate to devise a man-hours budget for TQM activity. One Water Authority's Engineering Services function adopted this approach which went a long way towards convincing hard pressed engineers and project managers that the company was serious and intended to allow them the resources to complete the job. Their project allocations were adjusted to provide time for the TQM/ISO9000 projects.

Attention to the objective setting, performance appraisal and reward systems is an essential element in communicating the need to spend time on the activities of quality improvement.

In summary, one can think of the implementation process in TQM as being much like the systems in a human body, or any other live organism. The brain is the equivalent of the understanding of the concepts of TQM by the executive management team. The education process is represented by the nervous system, carrying the concepts to all parts of the organism (organization). The shadow team structure provides the skeleton and the implementation plan the muscles by which the skeleton is moved.

CHECKLIST/IDEAS MENU FOR CHAPTER 5

Positive commitment

Publish a Quality Policy and have all your top team sign up to it.

Establish a clear mission for the organization and make sure it is simple, strategic and challenging. If you already have a mission, check it against the following criteria. Is it:

- Clear and unambiguous?
- Simple and understandable?
- Specific about exactly the business you are in?
- Specific as to where you wish to be in the future, in measurable terms?
- Focused on the central strategy rather than trying to cope with too many strands?
- Reflecting the distinctive competence (what you are best at) of the organization?
- Broad enough to allow flexibility in tactics?
- Helpful to managers in departmental goal-setting?
- Achievable?
- Motivational?

Define the values by which you wish to operate. Invite your key customers, local community leaders and employee representatives to participate in the discussions.

Close the plant and hold an open day, inviting your suppliers and customers along to hear how you are going to make quality a way of life.

Do a quick study of your cost of quality and post it on all noticeboards. Have a board made for the entrance showing your target for improvement of quality.

Declare your customer complaints figures and target for elimination.

Train your managers in TQM yourself if you are the CEO.

Hold some unplanned meetings with employees to discuss their problems over quality and the company's approach – in the canteen, the social club, the locker rooms, the car park, wherever you can.

Put quality at the top of every board agenda, and insist on other managers doing the same at their staff meetings.

Set up a **blame-box** at management meetings and have your managers put in a fee whenever they blame the person instead of addressing the process.

Stop the job when you have a quality problem and get everyone concerned involved in finding the cause.

Keep a checklist of all promises you make, no matter how small, e.g. I'll ring you back, I'll look into it, and monitor how many you fail to keep. Set a target to routinely keep all promises.

Find simple phrases and actions to highlight attention to quality, regularly.

Planning

Use the survey data you have produced as the starting point for focusing attention on priorities in TQM implementation.

Set up a steering group for the process; focus them on changing the cross-functional rather than departmental processes in the business.

Set up small planning groups among the steering team to help each other develop related action plans.

Set up review teams and establish Quality Policy Deployment as a management system.

Use the critical success factors, derived from the mission, to translate the latter into meaningful plans and actions at all levels of the organization.

Plan the shadow organization structure needed to secure maximum involvement from employees and support from middle management and supervision.

Early communication

Hold mass briefings with the workforce on your plans and the strategic focus on quality. Get a contribution from all your top team. If you are unionized, check out the attitude of officials and get them to participate if they are willing.

Make a corporate video on quality.

Set up a telephone hotline on the initiative.

Use the electronic messaging system to display the status of the programme.

Launch a regular newsletter; circulate articles; hold informal case study meetings on them rather than simply leave it to people to decide if they will read them or not.

Appoint a TQM communicator responsible for co-ordinating and disseminating information on quality.

Circulate quality problems, customer complaints and measurement data on them to all staff.

Use the team briefing process to communicate the plans for involvement and the actions being taken at the steering group.

Include an overview of TQM in your new employee induction programme.

Incorporate the quality policy in sales literature, logos, headed paper and other corporate projections.

Hold quality review meetings with key customers to find out their views on your performance and where you should be improving. If they are involved in TQM themselves, explore any possible joint events. Get customers to give presentations of their needs to key staff in your organization.

Put the quality manual and other information about the programme in reception.

Have plaques made of the policy and mission.

Hold a vendor quality day to educate suppliers in their own role in ensuring quality improvement is fed back down the supply chain.

REFERENCES

[1] Quoted in an article by James Creelman in *TQM Magazine*, published by IFS Publications October 1991

[2] From an article by Brian Parsons in *TQM Magazine*, August 1991

[3] Peters, Thomas J and Waterman Jr, Robert H (1982) *In Search of Excellence*, Harper and Row, New York

[4] Peters, T (1987) *Thriving on Chaos*, Pan Books in association with Macmillan, London

[5] From an article by Dr Sam Black in *TQM Magazine*, November 1988

[6] Quoted in Beckhard, R and Harris, R T (1977), *Organization Transitions: Managing Complex Change*, Addison Wesley, Reading Mass.

6 Getting organized

Participation

participate – to have a share, to take part in something

6.1 THE CID CONTINUUM

Participation is given a chapter all to itself because it is a paradox – we all want it, government, shareholders, management, unions, shop floor alike but we are all at our deepest roots scared stiff of it. It is not easy to do for a number of reasons. Firstly, we often fail to understand what we mean by participation. The dictionary definition is given above. Too many managers and government and industry worthies confuse it with communication approaches such as team briefings or corporate videos and journals. As we explained in Chapter 4, if one looks at the sophistication of the employment relationship in organizations, it seems to develop along a continuum moving from communication through involvement to development, what Richard Barnes calls the CID continuum. It is depicted in Figure 6.1.

In the 1960s and 1970s, many industrial organizations were operating well to the left. Even communication was being left to third parties such as shop stewards and works council representatives or else it was achieved through impersonal channels such as memos and noticeboards. One of the effects of the 1980s industrial relations revolution, led by Margaret Thatcher, was to return the responsibility for talking to people firmly back into the hands of managers. Renewed interest in methods such as team briefings and walking the job reinstated face to face communication as a key management skill.

6.1.1 Communication

The communication area of the continuum covers organizations on the extreme end which have hardly any mechanisms for communicating with employees, to organizations employing more sophisticated and systematic

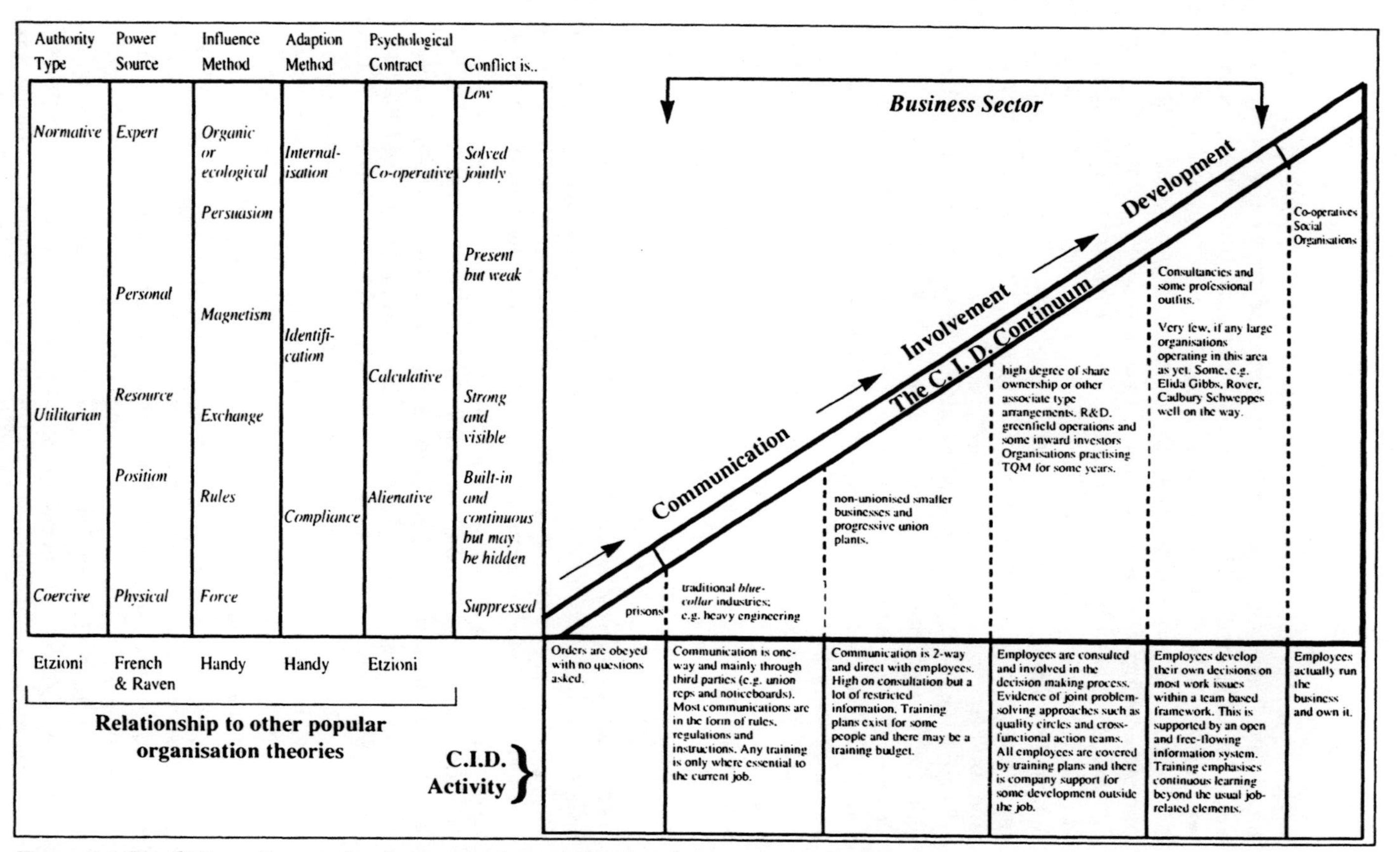

Figure 6.1 The CID continuum showing comparisons with related management theories.

methods such as team briefing, journals, business reporting, corporate videos, etc. These techniques however should be seen for what they are: communication media. They do not constitute involvement except in its very mildest form. Often forgotten is the need for upward as well as downward communication. IBM, for example, run a 'Speak Up' programme enabling employees to write in confidence on any matter which concerns them at work and get an answer from a senior manager within 10 working days. Analysing the letters gives valuable trend data to the company on actual or potential causes of discontent. This also ties in and enhances the information available through their regular employee attitude surveys.

The content of the communication also seems to be important. In the Industrial Society survey referred to later in this chapter, only 39% of respondents said they were interested in the company's financial performance. Highest rated information needs were plans for the future and how they are doing in their job. Only half of those interested actually got feedback on their performance. These results should not surprise us. The desire for information about future plans is related to job security – one of Maslow's lower order (i.e. basic) needs. Some employers' tendency to quote a set of figures at their workforce every month (profit, sales, output, etc.) without any interpretation of reasons for and meaning of the data, undoubtedly turns people off.

This fact probably explains why so many organizations find that the response of their staff to the introduction of team briefings is disappointing. The main problem experienced, even where there is a regular process established, is that communication upwards through the briefing system is poor. Obviously if management is talking to its staff in a foreign language they are unlikely to respond. Another difficulty is the degree to which briefings are **natural**; i.e. perceived as the usual way of doing business rather than a bolt-on system which is different from the kind of communication which usually takes place between managers and their staff. If you only ask people for their views once a month, this makes most people uncomfortable about giving them. Giving views may be seen as threatening (as indeed it is in many businesses where any criticism is seen as something arising from a mental deformity in the critic). If talking **with**, rather than at or to, employees is the daily norm, airing opinions at team meetings will be seen as nothing out of the ordinary. Unless team briefing as a system is supported by the behaviour of walking the job and the attitude that people's opinions matter all of the time rather than just once a month, the upward part of the briefing will remain weak and ineffective.

This psychological defence of silence to perceived threat is not restricted to the shop floor. Managers will react similarly if the style of the business is autocratic from the centre or the chief executive. One chief constable of a regional constabulary explained to us how he had started a regular series of conferences with his chief superintendents in order to increase the level of involvement in policy making. His first three events were met with nothing but nods of agreement and silence from his audience when asked for their views. Breaking the ice was one of the toughest things he ever had to do. Many would have given up in disgust and written the problem off to lack of interest. However, he persevered and eventually succeeded in generating a debate to the point where these reviews of policy are now a valued and hotly debated forum for two way communication between himself and his senior staff on issues that concerned them but on which they had passed up the opportunity for contribution for fear of the effect any views would have on their careers.

The way in which team briefers are trained is one of the causes of the problem. Traditionally, training on team briefings is treated as a matter of presentation skills, which is exactly what it should not be. Presentations rarely get reactions. Questions, suggestions, inviting ideas, talking on the same level and in the same language as the people you are talking with is the way to get reactions. We should train team briefers in interpersonal and group discussion skills rather than in the sort of techniques which are more at home in a sales pitch. And, believe us, employees are quite capable of distinguishing between a sales pitch and a genuine interest in exchanging information and ideas. Too many team briefs are in fact a thinly disguised piece of carefully vetted and one-sided propaganda, instead of a factual report and a basis for real bi-lateral communication.

This is partly because we live in a secretive and manipulative society, from government down through the judiciary and into our social and organizational life. We need to change our approach. One of the reasons for disputes and public hiatus over so called discoveries is that we hide so many things we simply create opportunities for leaks to arise. Information is like water, it should be allowed to flow naturally. The more dams we build in the river, the more leaks we are constantly having to plug, the more dam-busters there are around, the bigger the disaster when the dam is breached and the more policemen we have to employ to police the dams. This is a tremendous waste of energy. If as much management time were spent communicating as is spent on pontificating whether to or not, or else getting approval to release information, there would be far less unnecessary emotions released when bad news gets out. A company in the paper industry changed its owners from an established British business to a US parent. Prior to the change the British parent had a policy about salary information. Shop floor rates were to be posted on the noticeboards each year but staff rates were not to be released. They even insisted on a more senior person (does seniority correlate with trust?) handling the staff payroll. Senior managers' salary rates were even hidden from the local personnel manager, being kept under Corporate HQ control.

How much management time was spent trying to find out how much each other earned? How many hurt egos and raised hackles were there when people found out? How much internal politicking then developed to try to influence the status quo? How many computer hacks spent how many hours devising ways of tapping into the staff payroll computer? How much hot air was let loose when they succeeded?

When the US company took over, they brought in a new payment system. Once agreed and salary bands set, they published them all, from chief executive down, on the noticeboard. All of a sudden there was an increase in dam-buster redundancy.

Instead of having to justify the release of information, we need to justify any decision not to release it. The norm should be to tell, the exception not to.

6.1.2 Involvement

Organizations operating in the involvement span of the continuum have evolved beyond merely telling people what is going on, to actively seeking their contribution to (not just opinions on) the decision making process. Cell-based manufacturing, autonomous and semi-autonomous work groups, Quality Circles, joint planning and problem-solving forums are all forms of involvement. Good communications are a prerequisite for operating in this area of the continuum. Only a minority of organizations in industry could be truly said to be operating in this middle span, but the trend over the last 10 years has certainly been to move in this direction, often following the lead of inward investors such as the Japanese, Germans and Americans and the opportunities presented by reconstruction and greenfield sites.

The cultural effect being sought is a sense of **ownership** of the company among its employees. This can have remarkable effects on employees' commitment to the company and the type of activities they will undertake. In the distribution arm of Coca-Cola Schweppes Beverages we came across warehousemen with business cards who actively promoted the product in their own time at events where the Coke or Schweppes name was present, in sponsorship or just the provision of drinks.

The keys to involvement are several and complex. At one level they can be financial. Share ownership and profit distribution plans can help to foster an interest in a company's affairs at the competitive level which is often hard to get across in the normal day to day routines of workplace activity. Some evidence exists for suggesting that limited positive benefits can accrue from this approach. The Industrial Society survey referred to below showed that one in six UK employees own shares in their company. However, this was heavily skewed toward managers amongst whom one-third held stock. Amongst unskilled manual workers this dropped to just 5%. Half of all the share owners surveyed felt that owning shares had made them more committed to the company's success, although there was no significant difference on job satisfaction or their rating of their company overall as an employer.

Clearly, disposable incomes are a major factor in determining the spread of share ownership. Those on low and average earnings can scarcely afford the luxury of risk investment. If a company wishes to increase its employee share ownership an incentive is essential. One of the most common forms of incentive is the use of Employee Share Ownership Plans (ESOPs). These typically offer a one for one share purchase arrangement, with the company issuing matching shares from a share trust for every ordinary share purchased; effectively a half-price buying arrangement but with the added bonus of tax savings.

The amount of capital allocated to the trust varies. Recent figures showed that ESOPs ranged from 80% of the share capital in companies such as People's Provincial Buses to as little as 5% in MFI Furnishings. For many companies the size of the scheme may be determined by reasons other than philanthropy. ESOPs are a very powerful hedge against unwelcome takeovers.

Profit sharing schemes are an alternative or complementary approach, bringing the results of performance a little closer to the influence of those in the scheme. Unskilled manual workers in particular responded to the Industrial Society survey to the effect that being part of a profit sharing scheme would make them work harder.

Pay distribution systems such as performance-related pay and skill-based grading structures can also emphasize or detract from the attention to quality. Appraisal schemes which emphasize personal objectives may shift attention from team objectives and involvement in cross-functional activity unless such activity is part of the objectives set. Rewards for getting it right first time must be at least as visible as rewards for fixing and being Batman (See Chapter 5 – recognition and reward role on the steering committee).

Job security plays a significant part in ownership. Doubt as to whether you will still be with the company next week are hardly likely to encourage a sense of belonging! Again the Japanese have recognized this rather obvious truism for many years. There is some evidence that this approach is starting to impact on the organization of labour contracts in the West. The trend toward core and peripheral workforces is one signal which has been documented elsewhere. Rover's **new deal** in 1992 with its employees reversed an 80-year-old tradition in Western car manufacturing by giving its employees job security and lay off guarantees. IBM have always had a no compulsory redundancy policy, albeit this has been put under considerable strain in the 1980s and 1990s.

Even where TQM has been established in volatile industries, it has been accompanied at the least by a more cautious approach to redundancy and, if unavoidable, a greater emphasis on redeployment and post-contract support than previously the norm. Rank Xerox for example in their major downsizing (horrible non-word, that; what the hell's wrong with reduction?) of the 1980s provided redundant executives with preferred supplier contracts to help them establish their own businesses, guaranteeing Xerox as a customer for a period – a kind of new company umbilical cord.

The structure of the work organization itself determines many facets of the employment relationship; in particular, job design can influence the degree of control an employee has over his or her work and, with that, the degree of personal responsibility felt for the outcomes and quality of work. Traditional systems of shifts, absence coverage and in-line production methods tend to stifle personal accountability; the job will be done whether or not the employee turns up and nothing is left incomplete at the end of the day, unlike most managerial and professional jobs. Consequently employees constrained by such systems feel little sense of ownership of the total process.

Studies of organizations moving toward cell-based manufacturing and team structures in job design frequently show falling absenteeism, higher productivity and better quality of output as a result of the greater sense of involvement and decision-making responsibility which comes with these systems.

Digital Equipment[1], for example use a derivative of autonomous work groups called 'High Performance Work Design'. Groups consist of around 12 members with full responsibility for assembly, test, fault finding, problem solving and some maintenance. Each group had a leader to start with but now that group decision making has been established, the leader has withdrawn. Management takes a supportive role. The company feel they have achieved better quality, higher output, lower inventory and faster and more accurate decision making.

These changes are not just confined to the high tech end of manufacturing. Emcar, a clothing manufacturer, changed from traditional production lines to Autonomous Work Groups (AWGs), with individual piecework being replaced by a group bonus. As a result, absenteeism and labour turnover dropped to well below industry norms and productivity increased. Turnaround times on average orders were reduced from 6–8 weeks to just 4.

6.1.3 Development

Involvement, as we have defined it here, takes a knowledgeable and well informed workforce. The issue of training has to be addressed in any move up the CID continuum. Organizations which recognize that the benefits from good communication and involvement require the catalyst of continuous development of their employees' knowledge and skills will be seen operating in the development end of the spectrum. Sadly, very few organizations have yet understood the importance of development as an ongoing process rather than an occasional shot in the arm. For example, a 1989 survey[2] showed that only 27% of companies provided training for customer service staff. The then Training Agency, reported in its survey of training in Britain in 1987 that two-thirds of all employers in the UK had no training budget and three-quarters no training plan. 85% did no analysis of effectiveness or benefits.

Although this state improved with the assistance of huge amounts of publicity behind such drives as the Management Charter Initiative (MCI), the launch of Training and Enterprise Councils and the efforts of the DTI and DES, the importance of training to people is still understated. Mintel, in a 1990 survey, found that 20% of employees rated more training higher than more holidays or other fringe benefits. The preference was for in-house training, especially from women. In our own training needs surveys, we have often found a large pool of employees desperate to know more about the organization's business and the role of other functions within it; something easily arranged without great expense, yet frequently overlooked in the training plan, if one exists.

Some organizations are setting the pace in training and contrary to popular belief, not losing their best staff to competitors. Rather, they are building their knowledge base and attracting staff with potential who look for commitment to training as a benefit. Sheerness Steel, one of the most successful and profitable steel producers in the UK, had embarked at the time of writing this book on a policy to have 75% of their workforce with vocational or professional qualifications by 1992[3]. The training budget has risen six-fold, but at the same time, tonnage has almost doubled, and productivity levels are considered to be alongside those of their best German competitors.

Corning Inc. has increased employee training from 1% to 4.5% of its payroll in 5 years of TQM. Wallace Corporation's 5 top executives have had over 200 hours of formal training in quality improvement tools and concepts and its commitment to TQM is highlighted by the fact that at least one of the 5 has been involved in every training programme for its associates (employees). Motorola have a minimum of 40 hours of training for every employee each year[4].

Much of this effort is also in-house and focused on the specific needs of the business and its employees. The Macdonalds University in the USA is legendary. Elida Gibbs in the UK has a **Learning Centre** in the same vein. Ford, Lucas, Rover and other leading-edge businesses in their fields have all established programmes to develop a continuous learning process extending well beyond the usual job-related opportunities to which employees in most organizations are restricted.

Of course, development is not limited to just management or skills training. It is also about developing people on the job by creating **learning**

opportunities. When a problem occurs, a learning opportunity arises. This would be a welcome change for many employees who think that the only thing likely to arise is a boot!

Job rotation provides an opportunity to develop people's understanding of the organization as a whole. This should not be on a narrow basis among departments or production teams but on a wide basis, with employees moving between staff and line functions. Let production people spend time in product development, sales people in research, designers and architects on construction sites. Anywhere you have traditional conflicts and misunderstandings of requirements should be a candidate for the rotation programme. Most importantly, everyone should have the opportunity to see things from the customer's end of the tunnel. Massey Ferguson operate a product appreciation programme for employees at their Coventry tractor plant. They use a one day training programme and organize customer visits for their staff to see how the tractors are used in the field (*sic*) by farmers.

The CID continuum provides an interesting organization level parallel with Tannenbaum and Schmidt's[5] continuum of managerial behaviours and team maturity (Figure 6.2). This continuum is traditionally depicted as a model for a range of alternative management actions depending on the task to be achieved and the ability in terms of maturity and knowledge of the group led by the manager. However, it also helps to explain why so many participation initiatives seem to fail. If, instead of looking at Tannenbaum and Schmidt's continuum as a range of options, you look at it as a slide, you can see that the force of gravity would naturally tend to roll you back down to the left as soon as you relax your effort to climb the slope. Snakes and ladders might be an equally apposite analogy; it takes effort to climb a ladder – getting bitten by a snake only requires carelessness! These analogies are painful but true. Consider the following not atypical set of events:

Manager X decides he needs a new overtime rota. He considers how to achieve this and decides it is too risky to just leave it to the group; he determines a strategy based on an open discussion with his team, but having his preferred option up his sleeve, meaning to gently prod the group toward this solution as they discuss the options. He is now operating at least on the face of it at point 4 – involve (Figure 6.2).

During the meeting the group fail to see the light and appear to be strongly in favour of another approach. He is now forced to declare his opinion openly and put it into the group for debate. Snake number one has just bitten him and he has rolled back down to point 3 – consult.

The group find all sorts of reasons why they don't like it and the debate swings back to their preferred option. Snake number two. The manager rolls down to point 2 and starts to sell his strategy to the group. This is met with resistance by the team and an accusation that he is not listening to them. They have by now become quite attached to their suggestion and

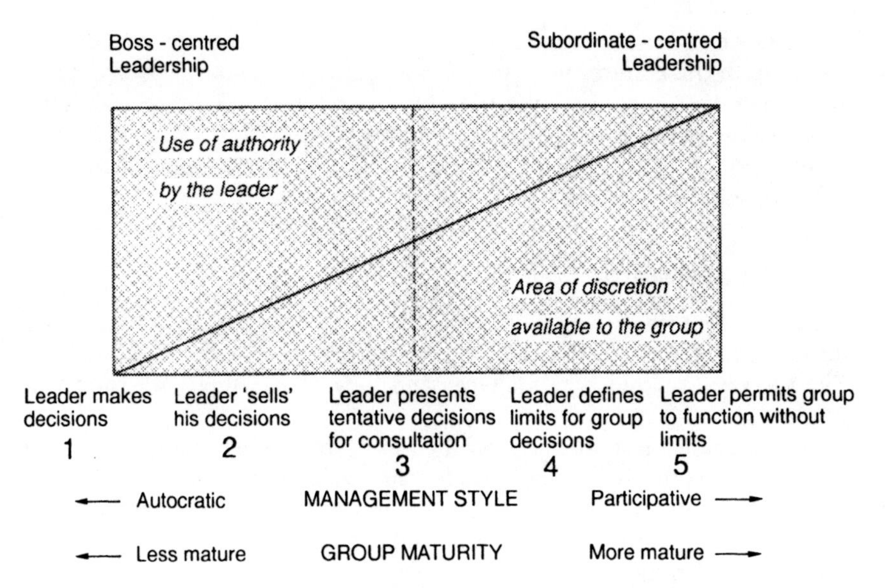

Figure 6.2 The Tannenbaum and Schmidt continuum.

don't see why they should give it up; after all '**you asked for our opinion**'.

Finally in desperation, the manager calls an end to the meeting and says he will consider their views and get back to them. The final snake has now got its fangs in his leg and he is back at the bottom of the slide. Next week he issues the new rota with a suitable form of words thanking people for their ideas and explaining why he has chosen the particular solution.

What is the result of this experience?

The manager wipes the sweat off his brow, determining not to listen to any more bloody academics (management trainers) again, reassures himself that people are just basically unreasonable, greedy, self-centred or whatever, and goes back to managing by dissemination.

The staff wonder why they went through all the bother of thinking up ideas when they are ignored and vow not to bother again.

When you come along with TQM, what effect will this experience, and thousands like them, have? (Refer to the hedgehogs in Chapter 4.)

6.2 CREATING THE ENVIRONMENT FOR PARTICIPATION

Participation is a matter of both dynamics and mechanics. It cannot be easily achieved because interpersonal dynamics are difficult to change. It requires systems and mechanisms to cause these dynamics to be changed.

Involvement will not occur because we wish it. It will only occur when we create both the culture, in terms of managerial behaviour, open, uncompromising listening, an error-friendly environment and organizational sub-systems such as work-teams, Quality Circles and cross-functional project groups which provide the opportunity for involvement within the system itself.

6.2.1 Management behaviour

We need to get rid of **daisy** management (**do as I say, you**) and root out and destroy the methods employed by these people for killing off good ideas. The weapons of the daisy manager and how to use them are shown in Figure 6.3.

Management must change from being the people who administer and deliver orders to being the people who encourage, coach, challenge assumptions about the constraints to improvement and develop their people to the point where they can make a difference to the status quo instead of defending it.

6.2.2 Creating an error-friendly environment

As we explained in Chapter 5, error-friendly doesn't equate to error-tolerant. Creating the right environment for quality improvement is a matter of planning (dealt with in Chapter 5) but it is also critical to the effectiveness of employee involvement in the organization for quality. Unless managers have a thorough understanding of the causes of error and under-performance, the tendency will always be to attribute blame for problems to those nearest to them – the employees. This is no way to secure participation. All managers must be educated in telling the difference between common and assignable causes and understand their own role in dealing with common causes. This will be discussed further in Chapter 7. Senior management must understand and communicate their understanding that they themselves have the responsibility for removal of problems in the system which prevent willing employees from doing a good job.

To support this, measurement must come to be seen as a tool to aid staff, rather than as an instrument to control them as it is generally used by most managers. This means that the targets are set by the staff themselves, within the limits of the process' capability. It is also important to cease measurement by output alone – put quality first is the message.

Implicit in this is that the top team, in the form of the Quality Steering Group, must be seen to be taking action on the process itself – with the involvement of appropriate employees in helping to identify the causes of organizational barriers and constraints. The team roles set out in Chapter 5

The boot – roll your eyes at the ceiling, grunt loudly, shake head despairingly. Put them firmly in their place. Stunned silence sometimes works well.

The bat – get them to go away and do a report on it which you can file away for ever or keep asking for further information, until it's their own fault nothing happened.

The sidestep – ask for the next contribution or say you will come back to it later, but make sure you don't have time.

Theft – ignore the idea but then come back to the next meeting with a slightly altered version and claim it as your own.

The history lesson – explain how we tried this in 1957 and it didn't work then.

The pedestal – embarrass them through overpraising the idea or crank the idea up into such a critical event for their own performance assessment that they find reasons not to risk trying it.

Paralysis by analysis – nitpick it from every angle until they give up.

The sub-committee – refer it to a sub group which you can stack full of awkward people who don't like the little upstart to begin with.

Mummy knows best – play on your superior status, science degree, length of service, World War 2 medal or anything else they haven't got.

Laughter – if all else fails, treat it with the contempt it deserves.

Figure 6.3 Management techniques for killing off good ideas.

provide one set of mechanisms for addressing relevant organizational systems and processes which may need to be changed. A bit of sackcloth and ashes doesn't go amiss either. Management must declare their awareness of their own contribution to organizational inertia and their intentions to do something about it.

All employees should be encouraged and trained in the techniques of constructive error identification. Treating errors as opportunities for improvement should become the norm. Rewarding and recognizing contributions to problem solving and improvement activity is the fuel which drives the continuous search for a better way.

6.2.3 Open, uncompromising listening

Few would argue that one of the keys to success in business is the ability to sense and interpret information from the marketplace. Who is closest to the marketplace? Answer – the employees. Salesmen, receptionists, customer service staff, committee clerks, telephonists, transport staff, drivers, production people all have an insight into customer perceptions; often a quite different and more detailed one than senior management, formal research or design people. They are the organization's antennae yet they are often underpaid, poorly rewarded and unrecognized. They frequently hold valuable information and clues to future actions which lie undiscovered unless management make a co-ordinated effort to release these trapped data.

In 1987, the Industrial Society gave MORI a brief to conduct a nationwide survey of employee attitudes which provided some startling and some expected but depressing results. Among its findings were that a half of unskilled manual workers felt that they could do more in their present job without too much effort. Those who said they could do more were less satisfied with their jobs, less involved in teamwork, less likely to rate their company above average as an employer and less interested in making a contribution to its success.

Only three out of five people interviewed felt able to express their ideas and contribute to solving problems in the workplace. A full 55% of manual grade supervisors fell into this category and most of them also felt poorly informed. The least well informed were most likely to be dissatisfied with their jobs and view the company as not well managed.

The same proportion of three out of five said they were usually told the reasons for what they were asked to do. The other 40% felt there was not enough opportunity to give their views to management. One-third believed top management was not in touch with what was really going on. Nearly one-half said that their management was more interested in its own views than listening. One-third of clerical, supervisory and manual employees believed they either had no opportunity to express their ideas or that no notice was taken of them if they did. Where suggestions were made, as many as one-third of respondents felt that nothing was ever done about them anyway.

One of the most telling results of the survey was that employees who were able to help solve problems were twice as likely to rate their company as well managed and be more committed to helping it be successful. Job satisfaction was highly correlated with teamwork, problem solving and feeling well-informed. Commitment was then most strongly related to job satisfaction and involvement and some form of profit-related pay.

The survey also challenged some old chestnuts about employee attitude to the job. Respondents chose interesting and enjoyable as the most important factors for them in a job. Basic pay came fourth after security and feeling that the job was worthwhile. Over 80% of manual employees felt committed to helping the company be successful.

An interesting link between the amount of information given to employees and the degree to which it was believed emerged: 90% of employees who felt well informed said that they usually believed management information. This dropped to 55% among those who felt they only got limited information. It therefore seems that if employees receive information infrequently, there is a tendency to be suspicious as to why they are being given it. Where open communications are the norm, information is taken at face value.

Amazingly, only 36% felt they were told the reasons behind major decisions and less than half of manual employees were given feedback on how they were doing in their job.

The most preferred method of receiving information was informally from the immediate boss, department/group meetings, briefing groups or pre-arranged meetings. The noticeboard was still the most frequent medium but preferred by only 17%. Least preferred was the grapevine and getting information via employee representatives (3% and 8% respectively).

Regular in-company employee attitude surveys are still a rarity although most organizations will have used them at some time or another. They have a value in the introduction of TQM, but again can only be a snapshot. To get the best out of them they have to be seen as a thermometer rather than a camera. Although not strictly a participation device, they are useful in giving early warnings of trouble, growing dissatisfaction with management or simply showing an interest in the opinions of staff.

6.2.4 Attitude survey design

For the purposes of this chapter, we will leave the wider use of surveys to others and consider attitude surveys simply from the perspective of their usefulness at the front end of TQM implementation.

One variation of an attitude survey is the culture survey described in Chapter 4. This however is aimed at employees' perceptions of the attitudes in the organization as a whole rather than their own views. A traditional attitude survey is rather more personally focused.

Surveys can provide a number of benefits. They demonstrate an interest in staff opinions and can provide a useful safety valve for release of feelings about the organization, in much the same way as a customer survey form at a hotel will allow one to vent one's dissatisfaction with a service which did not justify a more formal complaint. They may uncover grievances or concerns which were previously unknown but widely distributed.

From a managerial perspective, they may provide support for policy or employment systems changes and a range of suggestions for improvement of the quality of the employment relationship. If used with regularity, as at IBM, they can provide a valuable barometer of organizational climate.

IBM even uses employee attitude surveys as a measure of management quality, by creating a **morale index** from data provided by five key questions in the survey instrument. Each manager has his or her own score reflecting the opinions and attitudes of the people he or she manages.

The stages in the design of the survey are similar to those set out in relation to customer surveys in Chapter 4. The first consideration is a clear set of objectives and an acceptance that the results of the intervention will require some kind of response (see Chapter 4 example).

6.2.5 Objectives

Typical objectives for an attitude survey will cover areas such as:

- views toward co-operation from other departments or colleagues
- knowledge of the meaning of quality
- knowledge of and attitude toward the coming TQM initiative, if prior publicity has been given. Ongoing surveys may target views on the progress of the initiative.
- views of management competence and attitudes
- views of customers
- views on the extent and value of teamwork
- views on causes of problems
- views on adequacy of information received
- views on the extent to which they are involved in decision making
- views on how well trained they were for their positions and for future promotion
- views on quality of products or services produced by the organization
- views on value or appropriateness of pay and benefit systems
- views on social and welfare provision in the company
- views on the work environment
- satisfaction with job design
- suggestions for improvements

Whatever the objectives chosen, management must have in mind the areas in which they wish to see improvement and an idea of the measure of the improvement which is possible. Repeat surveys can show a measure of change, but bear in mind the expectation/satisfaction ratio explained in Chapter 4 and the impact of organization climate changes. For example, although you may have increased employee involvement activity between surveys, a recent event, such as a reorganization without adequate consultation, may have a disproportionate impact on perceptions of involvement among employees affected.

6.2.6 Timing

Timing of the survey is the next issue. If the information is to be used for planning the TQM implementation process, the survey should precede detailed actions. Equally, if the purpose is to measure the impact of TQM, then the survey should probably be preceded by information on the need for and principles of total quality, but again come before any visible activity, such as Quality Circles, takes place.

The adequacy of existing communication systems is another issue. As the Industrial Society survey mentioned, employees who feel well informed tend to have more faith in management information than those who do not. If communication systems are poor, there is likely to be greater suspicion among employees as to the purpose of the survey and the responses will reflect this suspicion.

6.2.7 Sampling

The next step is to decide who to survey. All employees may only be practicable in a small to medium sized enterprise. Similar principles to those concerned with customer surveys, of stratification, population ratios and random selection methods, must be observed.

If the introduction of TQM is being restricted to particular functions or is being phased in it may be necessary to limit the population to the departments affected. Beware however that when other sections of the staff are brought into the process, there is likely to be an expectation that their views will now be considered and the attitude survey repeated for them. One organization using a culture survey at the front end of a process which initially involved only a number of central support functions, decided to restrict the survey to just those functions. When it later brought its operational functions on stream, there were a good deal of rumblings among the new groups that their opinions were not considered as important as the **head office heroes**. Unwittingly, management had created another barrier between departments, instead of breaking them down, which a Total Quality approach is intended to do.

6.2.8 Question design and format

Rather than repeat the same points, refer to the key points in the customer survey section in Chapter 4. As regards content, you can afford a somewhat lengthier and more detailed document than for a customer survey since employees are likely to be more receptive to completing them than many customers. As a backstop though, questionnaires should be capable of completion within 20 minutes unless they are to be filled in under supervision or as part of an interview process.

6.2.9 Confidentiality

Confidentiality is paramount. Even if the survey is conducted by external agents, there must be verbal and visual assurance of confidentiality. Visual assurance could include placing any form into an envelope addressed, sealed and postage paid to the agent. Use of sealed envelopes and ballot boxes can also give confidence that the forms are not to be made available to simply anyone. If interviews by consultants with individuals or representative groups are to be used, the rationale for and process of selecting interviewees must be clearly explained. The consultants must ensure that each interview commences with an assurance of, and an exploration of any concerns about, confidentiality.

The design of any data sheets asking for information as to which groupings an employee falls into can be a source of potential breach of confidence. There may only be one female between the ages of 60 and 65 working in the finance function with over 25 years service! Data are essential to allow sub-group analysis but should not be so detailed as to give rise to concerns that the completer of a form can be identified from the personal data sheet. Remember also, if you are using a spreadsheet for analysis, that too much information may bring you into the realms of the Data Protection Act, if it becomes possible to trace individual employees' answers from the data.

6.2.10 Testing

Unless a recognized and well-tried format is being used, a pilot of the survey should be tried on a small sample of potential respondents before the final structure is agreed. This will allow potential ambiguities or misconceptions of questions to be ironed out before the survey has progressed too far. It may also be possible to discover people's feelings about the survey and their reactions to the user-friendliness of any questionnaires used. Did they for instance expect to be asked about certain issues which were omitted? If the questionnaire appears to only concern issues in which management might be interested, employees may feel frustrated that they have not had the opportunity to express their views on matters which really do concern them about the organization. The risk then is that some people will exert a certain amount of poetic licence on selected questions in order to try to get their point across. For instance they may give a low score on a question aimed at finding out if they have all the necessary tools to do the job, not because they are not provided with the tools, but because they are dissatisfied with communications on company progress and this question was the nearest thing on the form to that concern.

6.2.11 Communication

The survey's distribution should follow a detailed briefing of employees, preferably by their own managers. This should include the objectives of the survey, timescales, use of results, feedback to be provided, confidentiality and the process to be adopted for collection, analysis and disposal of data.

If necessary, a guidance sheet may accompany the questionnaire and possibly a briefing sheet about the purpose of the survey. In our own culture survey for example, we utilize a question and answer format to anticipate some of the more common concerns or misunderstandings about the nature of organization culture.

6.2.12 Response rate

Unless the survey is being conducted under supervised or interview conditions, a response rate of 50 to 75% is reasonable for an attitude survey. Obviously, the simpler the form is to complete, and the greater the desire of employees to express their views, the higher the response rate is likely to be.

Higher responses can also be encouraged by having time-out sessions for groups of staff to complete forms or by having lucky numbers with a prize draw, being careful of course that the confidentiality rules are not broken.

6.2.13 Carrying out the survey

The main points to watch for are the need for controls on distribution, completion and return of forms. Where interviews are used, the questions asked must be consistent; whilst not absolutely essential, the same person or persons should carry out all interviews. Each set of results should be traceable to a particular interviewer in order to spot any interviewer bias which might be present. Group interviews are useful in some circumstances, but can result in loss of confidentiality ('you know what old Fred said to her don't you?').

Once the survey is analysed and conclusions presented, an action plan should be drawn up to address the key issues and the results arranged in an understandable format for consumption by the employees. Again, presentations to staff of results by managers or the team who analysed the forms is preferable to posting the findings on the noticeboard.

Action plans should make use of project teams or departmental improvement efforts such as those discussed below.

6.3 ORGANIZATIONAL SUB-SYSTEMS

Various models exist for the creation of organizational sub-systems to translate the principles of quality improvement into practical results. Crosby advocates a departmental team approach, Juran the use of multi-disciplinary project teams, Deming and others the use of voluntary Quality Circles. All the approaches have one thing in common – identifying opportunities for improvement, in the form of problems or variation, and focusing small teams of people on resolving them with the aim of putting the solution in the hands of those closest to the problem.

6.3.1 The team approach in principle

At what point do a group of people become a team?

A group of people on a bus are obviously not a team. If the bus becomes trapped in a snowstorm for 24 hours without food, the chances are that that same group of people will become a team in order to collectively deal with a common challenge – survival. Survivors of disasters often meet up regularly again for years afterwards and may even give themselves names like the 3rd of October Club. Adoption of a name is one of the most powerful indicators of the existence of a team. (See examples mentioned later on cross-functional teams.)

Teams form one of the primary mechanisms by which objectives are achieved in organizations. They have a number of advantages over individuals:

- They contain greater collective knowledge
- They can enlist support faster and with greater commitment to goals
- They are often able to turn conflict into co-operation and problem solving
- They can tackle larger issues provided all relevant skills are present
- They generate better ideas through analysis
- They provide an important sense of belonging for most individuals
- They are able to share problems and reduce stress

However, teams can also be hampered by some potential drawbacks, particularly if senior management fails to fully comprehend the dynamics involved and fails to train their members effectively.

Teams are slower to produce results than individuals working alone. This is the price you pay for the extra knowledge and experience. They are harder to manage since the leader has to balance both individual and team objectives. Because of this, teams are sometimes considered a source of irritation and a reason for under-performance. However, it is usually the way teams are managed and trained, rather than the team itself which lies at the root of such problems. One other drawback to watch for is that they tend to take riskier decisions than individuals working alone. This may be because the blame can be spread if things go wrong or it may be that ideas are not challenged or understood properly if the prevailing culture is of the 'fly a kite' variety. 'Error-friendly' must not be taken to mean 'anything goes'!

6.3.2 Team dynamics and maintenance

The continuing performance of a team depends crucially on a number of factors:

- the continuing relevance and relative importance of the task
- the continued acceptance of the leader and the nature and style of meetings
- openness in communication between team members
- involvement of all members of the team and avoiding isolation of any individuals or their ideas
- proper maintenance and control of the meetings of team members with clear actions, objectives and responsibilities. These may be assisted by formal agendas and minutes.

6.3.3 Team roles and contributions

An effective team must not only contain an appropriate mix of skills, it should also provide a mix of personalities which complement the task at hand. Whilst personality clashes need to be avoided, the team must be able to disagree and argue constructively. Meredith Belbin[6] provides us with a useful model of team roles which can be used to examine the working of the team, team member selection and renewal (Figure 6.4).

The advantage of analysing the roles people play in teams through Belbin's typology is that it facilitates an objective assessment of individual strengths and views weaknesses as an allowable side effect of having a particular strength, rather than as something to be cured. For example, the completer/finisher may cause irritation to the group by endless repetition of minor points but his or her inclusion will ensure no loose ends are left unattended and will help keep the group focused on their objectives and promises. Individuals also find the self analysis useful in recognizing the effects of their personal contributions on other members and adjusting their style accordingly.

TYPE	TYPICAL FEATURES	POSITIVE QUALITIES	ALLOWABLE WEAKNESSES
Implementer	Conservative, dutiful, predictable	Organizing ability, practical common sense, hard working, self discipline	Lack of flexibility, unresponsiveness to unproven ideas.
Co-ordinator	Calm. self-confident, controlled	A capacity for treating and welcoming all potential contributors on their merits and without prejudice. A strong sense of objectives.	No more than ordinary in terms of intellect or creative ability.
Shaper	Highly strung, outgoing, dynamic	Drive and a readiness to challenge inertia, ineffectiveness, complacency or self-deception.	Proneness to provocation, irritation and impatience.
Plant	Individualistic, serious minded, unorthodox	Genius, imagination, intellect, knowledge.	Up in the clouds, disinclined to practical details or protocol.
Resource investigator	Extroverted, enthusiastic, curious, communicative	A capacity for contacting people and exploring anything new. An ability to respond to challenge.	Liable to lose interest once the initial fascination has passed.
Monitor evaluator	Sober, unemotional, prudent	Judgement, discretion, hard headedness.	Lacks inspiration or the ability to motivate others.
Team worker	Socially orientated. rather mild, sensitive	An ability to respond to people and to situations, and to promote team spirit.	Indecisiveness at moments of crisis.
Completer – Finisher	Painstaking, orderly, conscientious, anxious	A capacity for follow through, perfectionism.	A tendency to worry about small things. A reluctance to "let go".

Figure 6.4 Belbin's team roles.

6.4 TEAM STRUCTURES IN TQM

We can observe five types of team structure in a quality improvement process. These are:

1. A central guiding or steering team
2. Sub-groups of the steering team charged with particular parts of the implementation process
3. Consultative groups
4. Permanent, functional or departmental, teams
5. Cross-functional problem solving teams

The structure can be depicted graphically as in Figure 6.5.

The roles of the steering group and its satellite implementation teams were discussed in Chapter 5. These structures are primarily aimed at ensuring top management commitment by putting them firmly in the driving seat and securing the support of middle management through their involvement in the planning process.

The three other types of team are the basis of the **mechanics** of participation and are outlined in more detail below.

6.4.1 Consultative groups

One information sensing system is frequently missed in TQM implementation – user response to TQM itself. We tend to assume that everyone agrees with quality and therefore that they will accept what we are doing is right. Yet conversations with people on the sharp end of TQM often reveal a quite different picture. Sure, they have no argument with the concepts but the way it is put across, ambiguity and conflict in management messages and actions, the training programmes, methods, explanations and the other trappings of the implementation plan are less easily accepted. Some mechanism is needed to provide feedback to management on the implementation process itself and the opportunity for those on the sharp end to have some say in the decisions on such issues as training, methods and timing.

If the organization recognizes trade unions it will also wish to ensure that this does not become a barrier. Few unions have any desire to see quality programmes fail, but they cannot be expected to simply stand on the sidelines and watch their role being changed and their members' jobs being altered without any consultation. Ignoring the trade unions simply gives rise to suspicions that the central purpose of the exercise is to reduce their influence. We cannot recall any organization we have worked with citing this as a reason for adopting TQM. Indeed if it were, we would suggest that the process would be doomed to failure before it ever started. Having said that, the true reasons for TQM are unimportant.

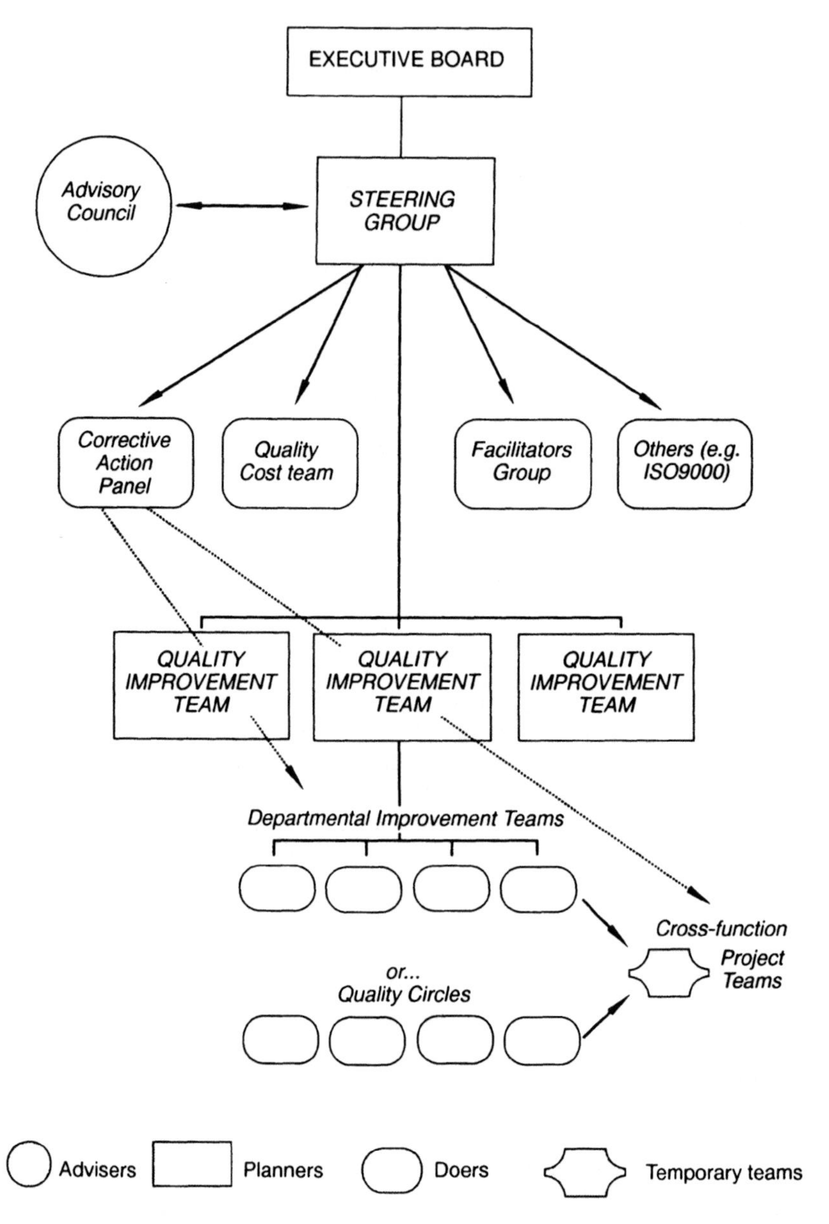

Figure 6.5 Organizational sub-systems in TQM implementation.

The same rule that Stew Leonard applies to customers pertains. Perceptions are more important than reality. If the perception is that it is a management ploy, then the barriers will be insurmountable.

Trade union stances toward TQM and other initiatives incorporating greater involvement of the workforce vary widely. Ford had immense difficulties in the early days. Certain parts of the Transport and General Workers Union have given the concepts stiff opposition, fearing the erosion of the traditional position of shop stewards. Certainly no-one can deny that their position is changed by such processes as TQM. More pragmatic approaches such as those of the Manufacturing Science and Finance Union and the Postal workers have accepted the inevitability of TQM whilst seeking to negotiate a different role, frequently more strategic, for themselves and their representatives and some form of reward for staff involved in higher level activities such as problem solving, process improvement, SPC and their accompanying skills increases. Trade unions may also seek to exert some influence over the selection of Quality Circle members and others involved in problem-solving activity. All of these approaches are quite rational and must be treated as such by managers, not written off in over-generalizations such as accusations of bloody-mindedness.

A Quality Council can often be a suitable vehicle for the involvement of trade union officials and other opinion leaders in the planning of TQM implementation and securing the views and suggestions from those at the sharp end on how the methods and activities are being received. Contrary to popular belief, it is rarely, if ever, trade union resistance which causes quality improvement initiatives to fail. The barriers usually go up much closer to home, often in middle management who potentially have the most to lose, hence the need for their early involvement in the planning process.

Other participants in consultative groups may be important suppliers and customers, particularly those with past or current experience of TQM implementation, shareholder representatives, community leaders and other interest groups at whom initiatives may be directed. A useful source of potential membership is to refer to the organization's own statement of values and beliefs and ask whether the parties mentioned in this document all have a line of communication into the development of the TQM process.

6.4.2 Permanent, functional or departmental, teams

There are broadly two approaches to the establishment of permanent Quality Improvement teams within departments or functions; the main differentiating factor is their voluntary or involuntary nature. Bill Jordan, the President of the Engineering Workers Union, on a DTI tour of Japanese industry, once asked a Japanese assembly line worker if she had volunteered to be a member of her Quality Circle or whether it was compulsory for her to attend. Despite the efforts of the interpreter, the concept of compulsory/voluntary could not be communicated. The assembly line worker replied that management had asked her to join the group; the thought of asking whether she had to go or not never entered her head. It was good for the company and thus good for her; how could anyone refuse?

Unfortunately (from the management perspective at least), this kind of common ground does not exist in most Western industries and therefore the question of voluntarism does have to be addressed. Quality Circles are by definition voluntary. There is even an Association of Quality Circles to promote their adoption and several books have been written about their use. The main drawbacks to a voluntary approach are that people can opt out of the process, sometimes giving rise to friction within work groups. In extreme cases, particularly where a confrontational industrial relations climate has prevailed in the past, a non-quality or anti-quality circle may form, often unbeknown to management, whose unwritten constitution is concerned with screwing up the activities of the Quality Circle, who themselves come to be seen as **management lackeys** or **the blue-eyed boys**. If subsequent promotions occur from within the group, it then becomes seen as a means for people to line their own pockets at the expense of **the lads**.

The alternative, non-voluntary, approach is for quality to simply become a regular meeting subject for the whole department who then select small sub-groups from amongst themselves to tackle particular areas of Quality Improvement. No-one is forced to contribute, but at the same time the structure of the group ensures that no-one can totally exclude themselves.

Whichever approach is chosen, some rules of operation are applicable. These can be summarized in five main areas:

1. Size
2. Constitution
3. Training
4. Focus of attention
5. Management involvement

Size

Group dynamics are such that size is an important factor. The advantages of greater knowledge and experience will tend to inflate numbers, whilst the objective of getting decisions made will tend to deflate them. Management must also avoid alienating those who feel they should be involved. Consensus may necessitate expanding the membership even at the expense of speedy decision making since a decision which is not supported by those affected will be ineffective, no matter how efficiently it was reached. Speed of implementation may require acceptance of a slower decision process; note the parallel with the prevention principle again – invest time up front to save time later. There is no absolutely ideal number but, as a rule of thumb our experience is that group sizes from 3 to 12 are suitable with an optimum membership of 4 to 6.

Constitution

The group should have established terms of reference. It is advisable for the steering group to have agreed a core set of terms which are flexible enough to be adapted at local level. Figure 6.6 shows a sample set of core terms of reference for a Quality Circle-type group, in this case known as a DIG (Departmental Improvement Group). Regular meetings should be scheduled, weekly on average and payment should be made if the meetings have to be outside normal hours. A team leader should be appointed, often the work group's own team leader or supervisor.

Training

Training is essential. Bunging a few people in a room together and telling them to come up with a list of things to improve/solve will result in the establishment of a knitting circle rather than a Quality Circle. Activity will be focused on pointing the finger at others, frequently management, for the problems experienced by the group. Team leaders should be trained in managing meetings, team dynamics and presentation skills, whilst all team members need training in problem-solving techniques and the principles of variation and process mapping and control. Some training in project management methods may also be of value, particularly where members may at some stage be leading a cross-functional problem solving group.

Focus of attention

Activity must be directed solely at things which the members of the group can do themselves. If outside resources are required these can be the subject of an objective justification, submission or presentation to management. The use of techniques such as Departmental Purpose Analysis (explained in Chapter 7) to establish requirements at each interface with customers and suppliers can be useful templates to guide the group toward activities within their own control. In the early days as the group is finding its feet, the team leader and possibly their management will need to define or approve the direction of pursuit in order to avoid the group going down

Departmental Improvement Group
Terms of Reference

Purpose

To improve the work and effectiveness of the department.
To create an environment conducive to quality improvement.

Membership

The membership of the DIG will comprise all members of the department headed by the Manager.
Sub groups may be formed to tackle specific improvement tasks.

Relationships

A Departmental Improvement Group will report, via its group leader, to the Steering Committee.
A DIG may call upon other departments for assistance on information gathering exercises to support its work.

Responsibilities

1. Define the role and duties of the chairperson, secretary and members.
2. Carry out Departmental Purpose Alignment and identify problem areas.
3. Recommend to the Steering Committee internal processes capable of improvement and action plans for priority areas.
4. Carry out process improvement plans approved by the S.C.
5. Raise Corrective Action Requests for improvements which are outside the department's control.
6. Monitor the effect of actions implemented and report to S.C.
7. Review the outcome of actions and identify further potential for improvement.
8. Help promote TQM by publicizing, internally, actions taken and successes achieved.

Meetings

Meetings will be held monthly as a minimum and more frequently if required.
Minutes will be taken by the secretary to document all decisions and actions.

Figure 6.6 Core terms of reference for a Quality Circle-type team. (Thanks to AMEC Design & Management Limited.)

blind alleys or taking on too much complexity before they are comfortable with the tools involved.

This is traditional group dynamics. A new group of people brought together with a common purpose become a team. In doing so they pass through four broadly recognizable phases of development; the rhyming memory jogger is **forming, storming, norming and performing**.

In the **forming** stage, the group are trained in problem solving and process modelling and control techniques. Leadership is provided by the team leader or supervisor, who carefully guides the team or selects areas for them to examine which he or she can ensure will lead to early success and understanding of the use of the tools. The group at this stage are probably happy to follow the leader and practise with the methods.

At the **storming** stage, individual members of the team start to reveal their personal agendas for change and may challenge the choice of activity by the leader, competing among themselves for issues to be addressed. This is a sign of increasing confidence. Weaknesses in the training of team leaders may be exposed, particularly inaccuracies in the application of the methods in use by the group. The leader must be able to present a rational basis for the selection of improvement objectives if the support of the group is to be maintained.

Coming out of the storming stage, the group resolve their differences and **norms** of behaviour and approach start to become established. These might include an insistence on sticking to the rules when brainstorming, voting on ideas or rating them rather then allowing the leader complete control. At this point, the team have developed their own ways of using the tools, adapting them to the processes and activity of their particular department. They are able to take on somewhat more ambitious suggestions and may increase their tendency to take risks. They may also now be starting to push at the barriers and constraints which exist – the **storm** may move from being an internal one to affect other parts of the organization.

Some teams find that personal identities for their teams help to establish a common sense of purpose. At CEGO, a manufacturer of building products and architectural hardware in Essex, teams have emerged with such names as **the Pussyfoots, the Asps**, (accounts, sales, production) **the Wasps**, (warehouse and stores problem solvers) and **the Pickle Pirates**. When the TSB set up a separate central mortgage processing facility supporting their bank branch network, each mortgage team was named after the surname of its principal internal customer, the regional manager.

The establishment of norms of behaviour and practice within the group is an essential stage in achieving cohesion between members and securing results from improvement activity. Thus they enter the **performing** stage and are now in control of their own destiny. The chair may be rotated; all members are equally comfortable with the tools and techniques and they naturally develop their own agendas for change. At this stage the only thing which can kill off their enthusiasm is lack of senior management interest (see below) or failure to respond properly to requests for resources.

This four-stage process of development has an apposite analogy with that of the rearing of a young hawk; at birth it is carefully fed by its parents until it learns to catch small things by itself. As it grows it needs less and less parental support to feed and progresses to larger prey, until ultimately it leaves the nest and can fend for itself.

Management involvement

Management must take an interest – not just in the first few meetings but forever. Most quality circle initiatives have failed in the UK – primarily due to lack of long-term management interest. After the first few meetings, the apologies for absence list at quality circle presentations gets longer and ever lower levels of deputy start to appear. The Quality Circle very soon gets the message that it is no longer important and meetings become fewer and fewer until it eventually withers and dies. Management interest is its lifeblood. Don't ration it.

Of course where Quality Circles have been instituted as a **good idea** without any strategic rationale, they were probably doomed from the beginning. The absence of any formal structure to ensure continuing management interest was a flaw of the system, not of individual managers. The steering committee provides a vital support mechanism for both the permanent teams and the ad-hoc project teams. In some organizations, each steering group member is assigned to a departmental improvement team and acts as its mentor. It is a good idea to insist that the mentors should not be linked to their own departments or functions.

Another alternative arrangement to departmental teams is to establish functional teams. For example, AMEC's Design and Management group at Stratford have set up functional groups of people who may work in different departments but have common processes. For example one team is made up of secretaries, whilst another is made up of planners and project engineers. Functional heads also have a separate improvement team. This approach has the advantage of overcoming the natural tendency for people to defend their departments against all-comers, since several departments are involved in each team, yet each team has common processes, such as managing diaries, or motivating line managers.

6.4.3 Cross-functional problem-solving teams

Many people, certainly those following Juran's approach closely, view these bodies as the engine room of quality improvement. Two of the most successful exponents of TQM in the United States, Milliken and Florida

Light and Power, had 20 309 and 1500 teams active, respectively, in 1989. Florida Light and Power reckon on a six to one return from their programme of targeted quality teams' improvements.

Certainly, for many organizations, cross-functional teams represent a radical departure from tradition. They appear under various guises; project teams, corrective action teams (CATs), Process Improvement Groups (PIGs) (the British animal loving nature permeates everything, doesn't it?), Quality Improvement Teams (QITs). The last one often causes confusion for some people – with Crosby's use of the same term to describe local steering groups.

Whatever their title, they share a number of features. Firstly they are temporary in nature, established with a single purpose in mind, disband when they have completed their task and involve staff from more than one department or function, with hierarchical status being of minimal importance. They are in fact a project team in its traditional sense, although it is surprising how many project management organizations struggle with making them effective!

They are probably the most powerful form of improvement mechanism because they can enlist the support of key managers, deal with a much greater part of the system and often have much less difficulty gaining resources than Quality Circles because the choice of project and membership is much more within the control of the senior management team. Again the steering committee provides an important link. A corrective action or project sub-committee may be set up to co-ordinate the setting up of teams, and ensure contact with them from among the senior management.

6.5 THE FAST-TRACK PROCESS

In the early stages of TQM implementation, we recommend that a number of fast-track projects be establish (what Juran would term bellwether projects). Implementing TQM has been described by Deming as like trying to turn a large oil tanker at sea. It cannot be done just by turning a wheel – the effects of such actions can take a very long time to come through. Likewise with TQM, the changes made by management and the education programme will not produce results overnight. In a large organization TQM may take as many as 3 to 5 years to really start to realize its true potential. As the initial interest generated by the training programmes begins to wear off, the absence of any significant early improvement can have a dampening effect on people's belief in, and management's commitment to, the process.

Inertia exists in many forms in an organization; attitudes, values, old ways of doing things, lack of knowledge, distrust, disbelief, seen it all before, etc. These things take time, a certain thickness of skin and constant repetition of the quality message by management, to overcome them. The fast-track process put a little oil on the wheels of change!

6.5.1 Objectives of the fast-track

The purpose of fast-track projects is to find a means of demonstrating the benefits of Total Quality in the short term. The education process itself, across a site, may take as much as a year to complete in some cases. Often employees come away from these training sessions without any immediate opportunity to try out the tools and techniques or see the results. If left too long, people will become disillusioned with TQM and will turn their attention elsewhere, believing it to have been just another management fad. Worse, management may assume employees are just not interested or capable of being motivated and allow the process to die, thus reinforcing any existing them and us attitudes. The fast-track project should therefore:

- be manageable by a small group
- cross functional boundaries
- be achievable in a short period
- challenge traditional thinking
- be visible
- relate to a long standing problem
- benefit employees as much as anyone else
- have enough scope to apply the 'new' tools of TQM
- affect one of the organization's Critical Success Factors.

We have often found that the technique of **Business Process Re-engineering (BPR)** provides a useful means of meeting the above objectives. The method is covered in more detail in the next chapter but basically involves breaking the organization into a series of horizontal processes such as order procurement, order completion, after-sales service, finance supply, etc. The most important processes can be derived from examination of the critical success factors and by looking for those processes which are failing to meet key market or business needs. The results of BPR can be spectacular and meet many of the criteria we defined because the technique invariably crosses functional boundaries, involves employees and is highly visible and different from the traditional ways of doing things.

6.5.2 Selection of the team

The team should be selected from among employees who have been exposed to the concepts of TQM and are part of departmental improvement groups or Quality Circles. If, as is likely, they have not yet completed training in tools and techniques, this should form the first part of their activity together with training in team dynamics.

6.5.3 Tackling the project

It is the process of problem solving which is as important as the task itself since this is the essence of the cultural change generated by TQM. The team should therefore commence its task by reminding themselves of the steps to effective problem solving and then follow the process through. Depending on the issue the process may look something like this:

1. Define the problem clearly, specifically and without implying cause
2. Identify data available on the problem
3. Define new data/information required to analyse the problem
4. Identify and implement any practical fixes to alleviate the problem in the short term
5. Build a flowchart of the process or processes involved
6. Identify the location of problems in the process
7. Brainstorm possible causes and draw a fishbone diagram (see Section 7.7.6)
8. Choose the most likely causes from the data collected or by voting/consensus
9. Establish suitable measures of performance
10. Define the steps in the project and assign responsibilities
11. Document and control each step and communicate results
12. Conduct a potential problem analysis before implementing any solution
13. Present proposals to management or the steering committee
14. Communicate changes to parties affected.

Management's role through all this is to:

- Remove the fear of failure; this doesn't just apply to the fear of failing with the solution but recognizing that fear may be created by simply establishing the project. An example in a large multinational company illustrates the point. After a seminar on TQM, the management team set up a fast-track project to look at ways to improve the planning process. The senior planner was invited to be a member of the team charged with the responsibility of reviewing the system.

  When the senior planner received his invitation, by memo, he promptly wrote out his resignation and handed it to his boss. Fortunately, deft use of a fishbone diagram by his superior was able to convince him that he was not the target of a witch hunt but a victim of the system.
- Provide resources and time to tackle the issue.
- Ensure training is sufficient to promote the success of the project. This will mean training in both problem solving and team/meeting process skills.
- Monitor progress, possibly through a senior coach or mentor to the team.
- Recognize and reward significant milestones.
- Publicize results.

6.6 SUGGESTION SCHEMES

People often ask us where suggestion schemes fit into a TQM programme. Some have concerns that they may conflict with the aim of getting people to submit ideas to their departmental or project improvement groups. This does not have to be so. The Japanese manage to run Quality Improvement teams and suggestion schemes side by side without any significant problems. This may have something to do with the way they run their systems. In the typical American or European suggestion scheme, only a few awards are made and these tend to be for significant identifiable savings. The award itself usually has some linkage to the cash generated by the idea. In a study in the USA, the average payout was $545. By comparison, the average award in Japanese schemes was just $3.

What our systems do is to single out a special few whilst the Japanese emphasize the contribution of everyone in the success of the business. Over 80% of Japanese ideas, averaging a savings value of $43, are implemented. As a result we end up with the enlightening anomaly that there are eight times more employees eligible to make suggestions to company schemes in the USA than in Japan, but Japanese companies receive 52 times the number of suggestions than those in the States. The position in the UK is not dissimilar. There are some exceptions; Rover Group's suggestion scheme receives 14 000 suggestions annually from 40 000 employees; a participation rate of 36% as against 5% for the UK nationally. The ideas saved them £4.5 million in 1989, with the scheme paying out £662 000[7].

The suggestion scheme introduced as part of Corning Inc.'s TQM process has received 10 times more suggestions than the old scheme and nearly half have been implemented; 50 times more than before[8].

This is still some way behind the Japanese. One plant of Nachi Fujikoshi, a machine tool and robot manufacturer, receives an average of 5.5 suggestions per worker per month, 70% of which are adopted. Nissan's Oppama plant receives 60 per year per worker.

The bottom line is that the returns from suggestion schemes in Japan are more than 13 times that of American and European schemes.

CHECKLIST/IDEAS MENU FOR CHAPTER 6

Communication

Do all parties mentioned in the statement of values have a line of communication into the development of the TQM process?

Review your policies on information release. Switch to a non-release by exception mode.

Launch a quality journal.

Set up a quality corner in the office with reports, before and after photographs and team awards.

Do your own video using real teams in action.

Stop dishing out meaningless financial statistics and concentrate on future business trends, new orders, processes, facilities and examples of successful employee efforts or brave failures.

Establish regular team briefings and get feedback from employees as to what they would like in them before you start.

Do de-briefings as a matter of course to ensure upward as well as downward communication.

Involvement

Set up a sub-group of the steering committee to look at ways to involve employees.

Consider the possibilities for cell-based as opposed to assembly line manufacturing processes.

Devolve power to the level of the team. Place recruitment, materials handling, job planning, measurement, etc., in the control of the team and its leaders through the use of autonomous and semi-autonomous work groups. Redeploy redundant supervisors and managers as process facilitators.

If you employ significant numbers of temporary, casual or agency employees, such as in the construction industry, consider setting up a team to look at ways of enhancing **belonging** among this group.

Review job design to look for ways to empower staff.

Let employees see, help design or choose equipment, software, furniture they are going to have to use.

Set up a Quality Council from trade union stewards and key opinion leaders. Consider involving customers and suppliers as well. Have them monitor and review the TQM implementation process and encourage constructive criticism and the occasional devil's advocacy.

Set up an employee share ownership plan covering all employees rather than the select few.

Consider introducing a profit sharing arrangement.

Ask people's views, especially those at the sharp end of the business in contact with customers – so obvious, but often overlooked, that we should repeat it a thousand times.

Involve middle management and union leaders from the outset in designing the implementation plans.

Run a culture survey.

Institute regular employee attitude surveys.

Are your letter signing authorities set too high? Why do you have them at all?

Do all people who could use business cards have them?

Discourage **daisy management** (see text).

Do not abandon the suggestion scheme. Review its use and policies to get many small ideas rather than a few big ones.

Development

Train all managers and supervisors in participative management, coaching and counselling skills.

Train teams in teamwork and group dynamics as well as problem-solving techniques. The process and interpersonal activities in groups are as important as the tasks themselves in achieving effective results.

Organize visits to customers by employees.

Establish an open-learning office at the workplace and stock it with books, magazines, articles, videos, company and customer material, process and product information, computers and self-teaching software.

Establish a budget and guaranteed level for training in the business; make sure all employees have a personal development plan and the opportunity to widen their education with company support. Ask the union to get involved if you recognize one.

Use job rotation to get staff into line positions for a while or put production people in front of customers.

Problem solving

Set up a cross-functional improvement project sub-committee of the steering group to establish key improvement projects and their teams.

Set up a blame box at management meetings and get everyone to put £1 in every time someone asks who did it instead of why did it happen.

Set up internal customer/supplier focus groups to learn about each others' processes and iron out interface problems.

Choose fast-track ideas to act as bellwether projects on quality improvement. Trawl for suggestions from employees. Don't forget to fast-track the team's training as well!

Set up quality circles or departmental improvement groups with a sponsor on the top team for each one.

Set up an award for helpful error-finding.

Institute a quality improvement prize for departments, Quality Circles or project teams.

Set up a quality hotline for employees to raise quality concerns. Use the electronic messaging system likewise.

Encourage name-making, myths and friendly competition among improvement groups.

Set up a cross-functional improvement project sub-committee of the steering group to establish key improvement projects and their teams.

Systems

Institute manager appraisal by subordinates.

Review the appraisal system for managers to question the degree of involvement they encourage among subordinates and their visibility to their staff (walking the job).

Examine your performance related pay systems and bonusing arrangements. Do they enhance or detract from quality and teamwork?

Evaluate all systems which encourage teamwork. Is it sought in interviews, emphasized in induction, training and performance reviews. Is it defined as a competence in assessment centres, management development or job evaluation?

Review your policies on redundancy and job security.

REFERENCES

[1] From an article by David Buchanan in '*Personnel Management*' the *Journal of the Institute of Personnel Management*; published in May 1987
[2] *Personnel Today* – 'Training in the UK,' (1989)
[3] Reported in '*Personnel Management*', the *Journal of the Institute of Personnel Management*; February 1990
[4] Extracted at various times from articles in *TQM Magazine*.
[5] Tannenbaum, R, and Schmidt, H W 'How to choose a leadership pattern',

Harvard Business Review, March–April 1958.
[6] Belbin, R M (1981) *Management Teams*, Heinemann, London
[7] From an article by H Gallacher, PERA, reported in *TQM Magazine*, June 1991
[8] From an article by Lyn Stolber, Quest Consulting, *TQM Magazine*

Managing the continuous improvement process

7

Process control
Problem identification

INTRODUCTION

The previous chapters have described the preparation phase for Total Quality Management, essential if the implementation phase, which is the subject of these last two chapters, is to be successful. Implementation of the initiative depends on careful planning, the build up of a firm and positive commitment from top management and the support, through active participation, of middle management and employees.

Despite careful planning quality initiatives can still fail. There are several reasons why many organizations fail to realize the potential of TQM during the implementation phase. Whilst the most frequent reported barriers are cultural change and middle management resistance, a serious weakness in many programmes in our experience is the shallowness of the training in, and use of, appropriate quality measurement and problem-solving techniques. This gives rise to a sponge-like phenomenon where the organization does some training and starts some projects but the existing culture gradually absorbs these efforts and reimposes its normal identity after a time. Organizations frequently report a flat point after about two years where progress ceases or goes into reverse. The seventh P for Permanence, discussed in the last chapter, therefore stresses the ongoing nature of improvement – not just in the projects or processes chosen but in the use of more sophisticated methods and growth in the skills of the workforce to apply them.

This chapter on managing the improvement process consists of the fourth and fifth Ps of quality:

Process control
Problem identification

Many quality problems occur because there is insufficient control of the processes in an organization. Lack of control of these processes and the failure of management to clearly define the requirements leaves employees to make their own decisions in isolation and in relation to the particular circumstances at the time. This means that there is great variation in the way that jobs are done, decisions are made, equipment is operated and maintained, and materials are purchased and used.

The amount of time required for problem elimination can be dramatically reduced simply by working to bring critical processes under control. This is why **P**rocess control precedes **P**roblem identification; if the process is in a state of control, i.e. predictable and consistent, there are likely to be fewer problems.

However, a process in a state of control does not mean a process which is acceptable; it simply means we know what it can do. It may be necessary to improve its performance (i.e. improve process capability). This is when the **P**roblem identification stage takes over. **P**roblem identification is treated as a separate issue from **P**roblem elimination since it is the stage where most problem-solving activity goes wrong. Unless a problem is correctly identified and defined, much activity is wasted in creating temporary fixes instead of eliminating root causes, chasing problems which do not exist or causes which have been incorrectly identified, or pursuing low priority matters which have caught management's attention by accident, snap decision or reflex action.

Problem elimination is a formal and purposeful process based on careful measurement and prioritization of quality problems, followed by properly resourced and planned corrective action with a view to the elimination of causes. The tools and techniques used to implement this purposeful process are covered in the final chapter.

7.1 PROCESS CONTROL

What do we mean by process control? Well, it is not the same as Quality Control although the latter may play a part in it. Process control is concerned with the management of processes in such a way as to ensure a consistent and reliable level of performance.

By **process** here, we are referring to any activity in which something is converted into something else by means of some activity, as depicted in Figure 7.1. It can therefore be a manufacturing process, a service or the communication of information.

7.1.1 Variation in processes

To understand how process control operates, it is firstly necessary to have some understanding of the principles of variation.

A variation may be defined as any departure from an expected outcome. This could be in traditional manufacturing areas such as tolerances around a target value as in materials or machined parts. Alternatively, variation may occur in the performance of people or processes or in standards and operating practices. In all cases the degree of variation lends itself to statistical study and analysis with the use of control charts.

7.2.1 Types of variation and their causes

The first stage in determining the appropriate technique to apply to any problem in performance is to identify the type of variation present; there are basically two, known as **sporadic** and **random** variation.

Sporadic variation is normally attributable to a specific and unusual error or change, such as a human mistake or a change in material. It will be a 'one-off' which can be traced and put right and hopefully followed through with a preventative action to eliminate the potential for that type of defect. Because it is a departure from the normal range of results, it is likely to have resulted from some specific change and have a clearly **assignable** cause. These are also known as **special** causes.

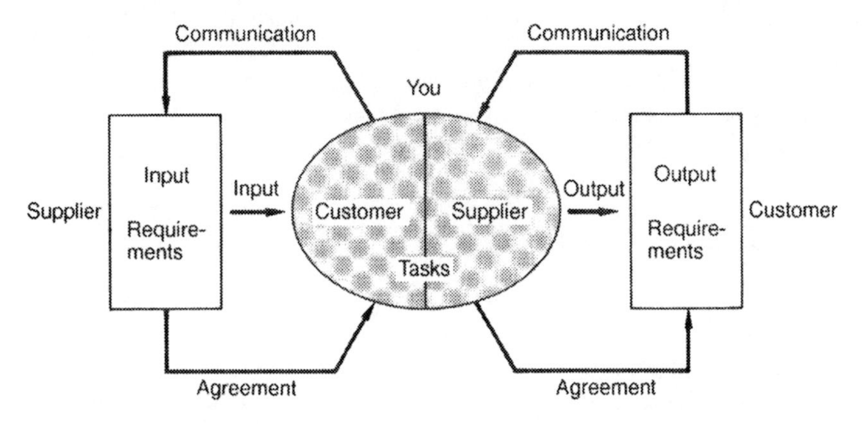

Figure 7.1 Process diagram.

Random or inherent variation requires much more effort to reduce it, but has potentially higher rewards. Firstly, because removing the causes of random or inherent variation is likely to have a wider effect than the removal of an individual defect, and secondly, because it may give the organization a competitive edge over its rivals. However, before we can study the causes of random variation, it is first necessary to remove all assignable causes as these will distort the data and lead to misinterpretation. The causes of random variation may be in the properties of the process or materials, or they may lie in the environment or conditions governing the operation. Either way they are distinguishable from the assignable causes in that they are naturally present in the design of the process itself and produce random variations rather than specific variations traceable to a particular failing. For this reason, they are known as **common** causes, since they are common to the process itself and not the **fault** of a particular person or piece of material or equipment. Generally the solution will lie outside the control of operational staff and will require some action on management's part to change the system governing the process. In manufacturing, for example, an assignable cause of a flaw in goods produced may be an operator incorrectly setting a grinding tool. Random variations would be those flaws produced by the natural tolerance of the machine, variations in temperature, raw material imperfections, etc.

In administrative processes, it is also possible to carry out this analysis. For example, an assignable cause of document errors might be a new employee brought in without the normal training having taken place. Random variation would arise from poor document format design, drafting quality, the training process itself, ambiguity of instructions, photocopying flaws, typing errors and collating failures, etc.

7.1.3 Dealing with assignable causes

Often, assignable causes are obvious and the temptation is to put them right and then dismiss them with phrases such as 'to err is human', 'no use crying over spilt milk', 'mistakes will happen', etc., or worse, treating the issue as a matter of discipline. True corrective action however requires that we do more than 'fix' the variation. Many common causes get treated as if they were assignable. In such cases, the true root causes remain undiscovered and will simply happen again unless the **process** which allowed the error to occur is tackled.

In a train crash in 1990, the driver was held to blame for over-shooting a red light. This was held to be an assignable cause (i.e. the driver's fault) and the driver was jailed. However the **root** cause lay in the system which allowed a driver to repetitively and routinely override warning signals. Unless the root cause is tackled the likelihood is that this **common** cause will manifest itself again. Techniques such as the fishbone diagram (see Section 8.2.6) are very useful in tracing back the root cause of such variations.

Actions to eliminate assignable causes usually involve some change to the overall process to remove the potential for that type of mistake (not by adjusting the operating process - see below) or, as a last resort, the introduction of some kind of checking or warning mechanism. Frequent occurrences of assignable causes point to the need for a complete overhaul of the process through mechanisms such as the process design review.

Adjusting the operating process, however, such as an operator moving the setting on a machine to compensate for what are in fact common causes of variation in the equipment or material, will only serve to increase the range of variation.

7.1.4 Dealing with random variation

In formal problem solving processes, it is typical to describe the process as passing through two stages; the **diagnostic** journey and the **remedial** journey. When dealing with assignable causes the diagnostic journey is often relatively short and most emphasis is on the remedial journey; i.e. the actions necessary firstly to fix the problem and then to eliminate the root cause. When dealing with random variation however the emphasis is on the diagnostic journey, since it is usually necessary to isolate a range of variables which may have an impact on observed performance, and these need to be studied in depth and subjected to experimental changes before the remedial journey is embarked upon.

Random variation is tackled in four stages:

1. Study of symptoms. This requires clear problem definition without assuming cause. Other means of studying symptoms include pareto display, autopsies, check sheets.
2. Hypothesizing as to cause is commonly achieved through brainstorming and cause and effect charts. Other techniques include structure trees, multi-vari diagram, theory tabulation, affinity diagram. Juran gives a useful summary of quality improvement tools in the *Quality Control Handbook*[1].
3. Data collection and display using methods such as control charts, histograms, x/y graphs, etc.
4. Data analysis using tools such as pareto, capability indices, flow diagrams, scatter plots and matrices.

Once assignable causes have been removed or reduced to the minimum, random variation is frequently left unattended, unless the level of variation transgresses outside the range of acceptable values. This typical management response, sometimes called complacency or, as Phil Crosby puts it, **the executive comfort zone**, can be seen not just in acceptable quality levels on products but also on administrative issues such as sickness absence, accidents, shipping errors. Usually these random variations require the application of what Joseph Juran terms a **breakthrough sequence**. This sequence is described fully under Problem elimination.

The most frequent mistake made by management is to interpret common causes of variation as assignable, usually by **assigning** the blame for the problem on the operator – a classic case of blaming the victim. London Transport has for all its history had fires in its underground tunnels. In the old days, steam trains were a major cause of fires on surface lines but since the days of the all electric service, regular outbreaks of fires, or as the company euphemistically called them, **smoulders**, continued to occur. The response of management was usually to treat each occurrence individually by taking appropriate action with the local station manager, permanent way inspector or contractor, urging them to regularly clear rubbish from tunnels and check work completion. They regarded fires as inevitable and in most cases down to human failings. What was not properly recognized, until the problem was dramatically brought home by the Kings Cross Station fire in 1987, which claimed 31 lives, was that over the years they had established a very successful and reliable **system** for producing tunnel fires. Statistical study of the frequency of outbreaks would probably have shown this system to be in a very good state of statistical control. Elimination of the causes required detailed study and analysis with the aid of outside expertise.

Exhortation rarely achieved anything, yet it is probably the single most used form of management action taken to deal with problems. This fact alone demonstrates the depth of our failure to understand (or be taught – our ignorance is part of the system!) the principles and lessons to be drawn from the study of variation in all types of process.

Later action by London Underground, such as the introduction of a ban on smoking, effectively changed the system and thus reduced the chronic or inherent level of fires.

7.1.5 Process Design Reviews

Deming insists from his work on variation over more than half a century that 94% of all organizational problems result from random variation in the system; in other words it is management's lack of understanding and control of the processes, people, equipment, environment and material in the system which requires attention in order to resolve the overwhelming majority of problems. Juran says over 80% of errors are traceable to management – Phil Crosby says it is probably 95% and in his more provocative moments has revised this to 100%.

Business Process Management is therefore critically important for the control of quality in terms of reliability and consistency. The Process Design Review, as the term implies, is a technique for reviewing and improving the functions and outputs of a process. It can be initiated for a number of reasons:

1. **Prior to the introduction of a new system or a change in an existing system.** This is a form of prevention to ensure that all the effects from a new or changed process are anticipated and planned for and all those who need to be aware of, or react to the change, are informed.
2. **To monitor a new or changed system or procedure.** If the process is complex or several parties are involved, regular Process Design Reviews may be built into the implementation programme. The introduction of TQM typically involves such reviews of the plan.
3. **To react to some change in the environment in which the process operates.** This might be competitor activity, such as a rival credit card company selling an ability to turn round applications quicker, customer demand for things such as additional information or features, or internal changes in ownership or organization structure. Such changes usually affect the requirements of the process and thus its functioning. Benchmarking studies, described under Problem identification, are also a source of stimulus for a review of a process.
4. **Repeated sporadic variation or departures from one or more of the requirements.** This is a reactive response to a problem which has reached an unacceptable level of pain for the organization or its customers.
5. **As part of a Quality Improvement Strategy.** Fast-track projects, Quality Circles and cross-functional action teams apply the technique to achieve a breakthrough in chronic or random variation in the functioning of processes. This is seen as a competitive strategy to reduce costs or gain an edge over business rivals.

Depending on the type of process, the review may make use of a range of analytical and quality improvement techniques such as:

Flowcharts
Matrices
Fishbone diagrams (see Section 7.7.6)
Process models
Focus groups
Audits
Customer (internal or external) surveys
Plan of Control
Process Capability Studies using control charts

Mapping the process can give useful information about sources of variation which may lend themselves to statistical study or design modification. A simple example is given in Figure 7.2 (a). The symbols used can vary; in this case the protocols as shown in Figure 7.2 (b) are used.

The stages in constructing a flowchart are:

1. Define the process or activity under review: e.g. **processing customer complaints**.
2. Determine the scope of the process: e.g. **from incoming letter to initial outgoing reply**. This step is necessary to ensure that what may seem a straightforward process doesn't get too unwieldy. In the example used, customer complaints may come into several points and in several different forms; telephone, letter, oral, etc. If we were to try to chart them all the diagram would be over complex and confusing.
3. List on a flipchart or piece of paper all steps in the process. Brainstorming may help here. If the process is not completely obvious, a physical study may be necessary. All steps must be

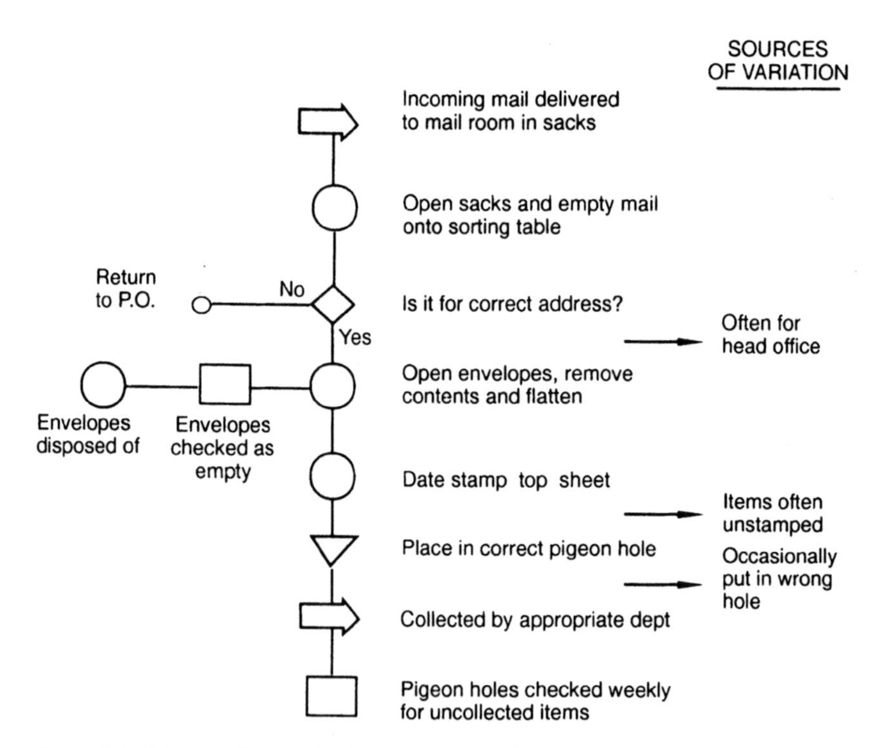

Figure 7.2 (a) Example of a simple process flow diagram.

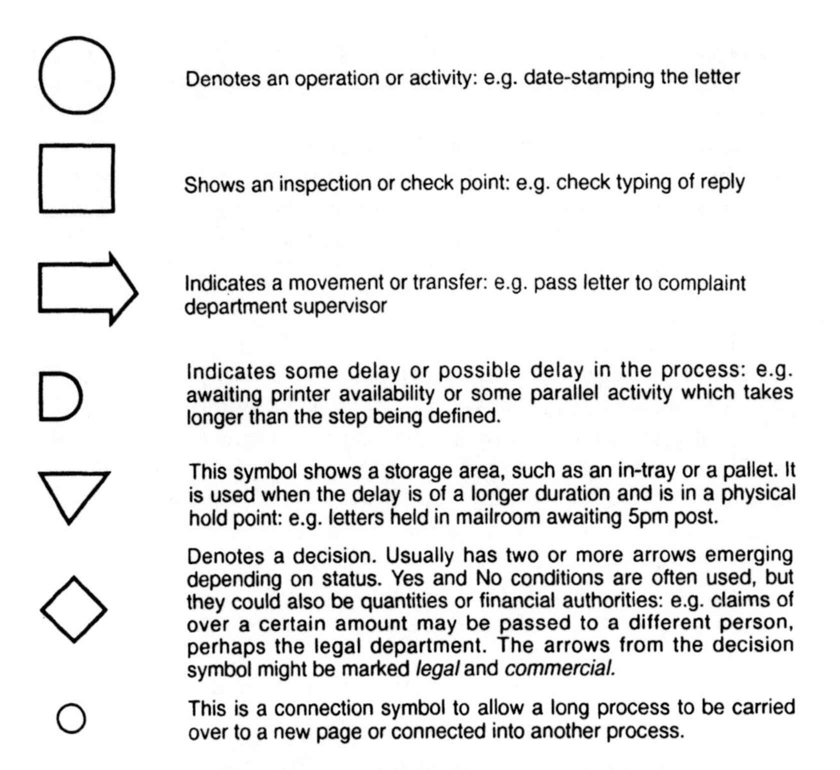

Figure 7.2 (b) Process flow diagram symbols.

included, no matter how trivial or fleeting some may seem. Put a symbol from the list against each step; check all symbols are present; if not you may have missed steps like transferring documents or checking.

4. Study the list and decide the broad shape of the flow diagram. Put the start point at the top or the left of the page as appropriate. Draw the flowchart and carefully describe the step within the symbol or outside of it if there is insufficient space. Put initials of the person responsible against the step if possible.
5. Continue adding steps until the chart is finished. Use connector symbols if you run out of space or wish to show that the next step becomes part of another process which is not under consideration; e.g. if a complaint is passed to finance to process a compensation payment.
6. Times can be added to the diagram if this is an important part of the study; for example, if the team's task is to reduce lead or response times. The diagram can be developed into a network

diagram if necessary in order to carry out critical path analysis. Another option is to cost each step if savings are part of the team's remit.

7. Identify any significant or regular areas of non-conformance by a red arrow and commentary. These may lend themselves to be control points for the introduction of control charts, process design changes to prevent the problem or inspection stages.

7.1.6 Using historical information

Historical information often provides a useful reference source for reviewing the effectiveness of a process; i.e. by extracting information about past problems or complaints, which were at the time dealt with on an ad-hoc basis. For example, Philips studied customer complaints in one of their large manufacturing plants. They found that 70% of complaints were to do with administrative handling rather than with the product itself; i.e. wrong date, number of parts, type of equipment, address, name, application, packaging, invoicing, etc. This led them to review the processes governing the way in which products were delivered rather than concentrating on the manufacturing process.

Reviews of historical data allow an organization to ask some broad questions:

- When was the problem first observed?
- What was the earliest stage at which the problem could have been discovered?
- Why was the problem not discovered at that earlier stage?
- What could have been done to have found the problem earlier or prevented it?

These questions often uncover some failing in the process such as the absence of auditing or monitoring and feedback mechanisms.

7.1.7 Process certification and ownership

In IBM, a system of certification of key business processes is used. These cover issues such as contract management, billing, accounts receivable, order entry, backlog control, software usage, inventory management, accounting, new employee induction, for example.

To apply the certification concept, an **owner** is designated for each key business process. For a functional or vertical process, the function head is nominated. In a horizontal or cross-functional process, one of the key functional managers is designated as owner. The key processes are then rated on a scale of 1 to 5 for features such as:

- Control of the process
- Adequacy of quality management
- Progress on quality improvement

This process of certification can be usefully interfaced with the **benchmarking** system which compares performance in the process against the **best in class**, which may not necessarily be in the same industry. Examples are given later in this chapter.

7.1.8 Process controllability

Process Design Reviews often uncover important gaps in the controllability of the process where requirements are unclear or ambiguous or where there is no means of monitoring, measurement or feedback. One of the major jobs of management is to ensure that the processes within which operators and administrators have to work are capable of meeting the agreed standards.

7.1.9 Quality improvement

The Process Design Review can be a useful tool in Quality Improvement in an area subject to criticism or complaint since it allows the parties to consider the wider process rather than the narrow symptoms. In one instance, an apparatus manufacturer used a large amount of steel and had to store a substantial proportion of his stock in the open due to a shortage of covered area. Although the material was treated with rust preservative, some of it still gathered rust leading to frequent complaints to the purchasing and material management departments. These departments reacted by blaming the space problem.

A Process Design Review identified the real problem as the lack of a mechanism in the process of handling steel to ensure stock was turned round on a first-in – first-out basis so that the rust proofing did not wear off before the stock was used. The team developed a process by which colours were added to the preservative to identify the period of application so that stock handlers could ensure they used all the steel of a particular colour before moving to the next batch.

7.1.10 Departmental Purpose Analysis (DPA)

Sometimes called **alignment** instead of analysis, this exercise goes under many different guises and is in fact a form of Process Design Review. It is however used as a specific project at the front end of implementation plans for TQM and follows the principle of putting preventive process control ahead of problem elimination in the stages of establishing a Total Quality strategy. The steering group member responsible for Business Process Improvement or customer care is usually given charge of designing and setting up the process, emphasizing the importance of the internal customer/supplier chain in meeting end user requirements.

In this approach, a systematic analysis of the functioning of a process, across departmental boundaries, is employed to produce a quality plan for the process. The plan identifies internal customers and suppliers and the various requirements for the outputs produced at each stage of the process, including those which have to be fed back to earlier stages of the process. The technique is also a valuable opportunity to obtain early employee participation in the improvement process by tapping their detailed knowledge of the process under examination and uncovering requirements which were perhaps hidden or felt to be too trivial to raise formally before.

Two examples from a personnel department of a major UK paper manufacturer illustrate two types of potential benefit from the process. During an off-site team review of query handling, the personnel manager discovered that his secretary had a regular source of variation in query answering caused by his failure to tell her where he was going when leaving the office. He was not aware of the problem; she did not think it important enough to make an issue out of; the customers suffered entirely avoidable delayed responses. A process was established to ensure this aspect of office procedures was covered in future.

In another example, new starter forms were completed using a triple-carbon copy form, which was photocopied twice, with one photocopy being retained by the personnel office and the local wages office respectively. The triple-copy form was then sent to the pensions department at head office. Some days later two copies of the form would be returned by head office and these would replace the photocopies originally taken in the personnel and local wages office, the latter being disposed of. **'Why do we send all three to head office when they only keep one?'** the analysis caused us to ask. Stunned silence; nobody knew. We rang head office; no-one knew there either. Apparently this office was the only one to send the pensions office all three copies. Eventually a former employee of the pensions department recalled that several years ago, this particular local office had had repeated problems with incorrect National Insurance data being put on new starter forms. This resulted in inconsistent data on the local personnel office computer and the one in pensions. To overcome the problem they had established a rule that all three copies of the starter form would be sent to pensions who would check the NI codes and send the copy forms back with any changes. These copies would then become the local master copies. Five years ago, the introduction of a new networked computer system had obviated the need for this process but it had continued to be carried out religiously, because no-one knew why they were doing it.

One employer related to us how they had rewired their starter motors on food processing equipment some years ago so that only one button needed to be pressed instead of a sequenced bank of eight which had been necessary with the old system. However, the operators to that day still started the machines by pressing all eight buttons, even though seven were blanks.

Simple improvements can also be identified; a substantial proportion of the Amazon rain forest was probably saved by a bright eyed employee in a data processing organization asking why the error check print out consisted

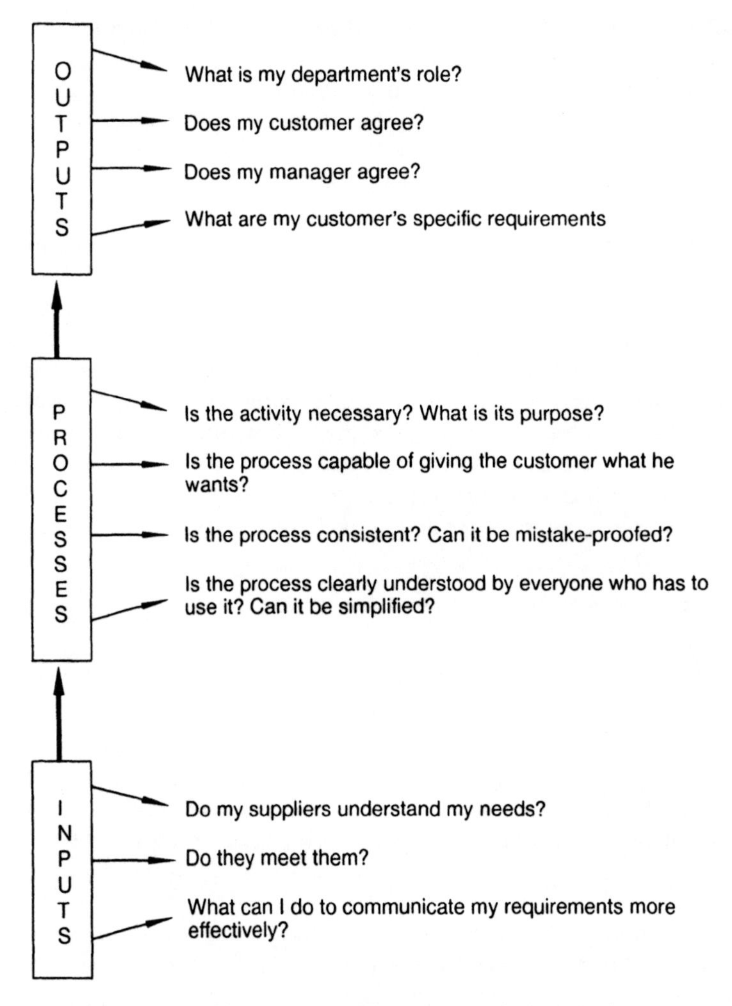

Figure 7.3 The process of Departmental Purpose Analysis (DPA).

of every single entry made in the shift, when the only lines the checkers read on the several thousand pages of print-out each day were those with an error message in the last column. A simple programme change reduced the number of sheets per print out down from thousands to less than ten. Apart from the rain forest, considerable time, eye strain and missed error line entries were saved.

The process of Departmental Purpose Analysis then, can both uncover needed requirements which have been suppressed, identify improvements and simplification steps and expose moribund activities. The skills of a monastic surplus materials disposals officer are required: get rid of old habits!

The pertinent questions to be asked during a Departmental Purpose Analysis are shown in Figure 7.3.

By gathering the information and analysing it within departments and across boundaries, DPA enables systematic improvements to be planned by each department to the quality of their own outputs and their relationship with supplying functions. Key performance criteria, from the customer's perspective should be established and appropriate measures of performance put in place to allow for continuous improvement to take place.

One of the side-effects of this process is that it allows internal customers and suppliers the opportunity to talk to each other about their needs and the things which stop them from being met. This helps break down barriers between departments, improves internal communication, encourages learning by staff of the process beyond their immediate jobs and broadens employees' perspectives and horizons. Staff interchange is a useful experiential technique which can support the DPA process and help to break down the tendency to negotiate rather than look for mutually supportive actions.

7.1.11 Business Process Re-engineering

Sometimes though, tinkering with the process and even major improvements are simply not enough. This is where the concept of Business Process **Re-engineering** (BPR) applies. The Process Design Review, although aimed at fundamental change, still relies on an assumption that the process itself is necessary. **Re-engineering** challenges the very *raison d'être* of the process. It is most useful whenever improvements of an order of 100% or more are required in a process, either because the process itself performs so badly or because a major competitive edge can be achieved.

Mike Hammer and James Champy in their book *Re-engineering the Corporation*[2] use a lovely analogy which we are pleased to paraphrase. If a business were a car and the management the driver, there would be three kinds of driver using BPR:

Driver 1 can be described as Desperate Dan. Desperate Dan has run his car into a wall. He gets out and looks at the wreck – he is in crisis. He uses BPR to get himself out of the mess and back on the road.

Driver 2 is called Prudence. She is a careful driver and keeps a good eye on the road. She sees a wall up ahead and stops the car. She gets her AA road map out and finds an alternative route which takes her round the wall, even if it takes her a little longer – she stays in the race but drops back.

Driver 3 is Vim Visionary. He is driving out in front, clear of all obstructions. He can see some of his competition in the rear view mirror. He decides this would be a good place to stop and build a wall . . .

Seven steps

There are seven steps to business process re-engineering:

1. Map the major processes which make up the business.
2. Define the **key** processes which have major implications for the organization's critical success factors.
3. Establish teams to review the process.
4. Map the process as it is now.
5. Go back to basics and describe the process as it could be if this were a greenfield operation.
6. Adjust the process to take account of unavoidable internal or external requirements.
7. Implement the new process.

Taking each of these in turn:

1. Map the major processes which make up the business

Most of our organizations are structured according to long-standing military principles of command and control. This means that strategy is developed from the top down and functions multiply and divide vertically below the leader, giving rise to a number of negative effects such as:

- Distance from the customer – functions see dealing with the customer as somebody else's job.
- Empire building – if success is measured by the degree of control, larger empires mean more success.
- Barrier building – the easiest way to **look** successful is to make others look less so.
- Sub-optimization of goals – set for functional rather than business reasons (e.g. Personnel concentrating on writing job descriptions instead of developing language skills).

- Tortuous decision-making routes up and down the command chain.
- Sporadic communication between functions confined to formal **events**.
- Unnecessary procedures, duplication and waste are not spotted.
- Some necessary actions and responsibilities fall through the gaps ('*I thought Accounts were responsible for that*').

Taken from a business rather than a command viewpoint, though, our organizations are in fact a series of **processes** in which one thing is transformed into something else – a customer order into a delivered product, a problem into a solution, cash into plant and equipment, unskilled staff into skilled staff. Taken from this process viewpoint, most organizations look remarkably similar. The cry '**but we are different**' starts to look a bit weak once the functional view is replaced by the process view.

The trouble with processes, though, is that they operate **horizontally** across an organization whereas our command structures are vertical. Management of the horizontal process is achieved by lobbing the end result of each function's actions on the process over the wall into the back yard of the next function.

The purpose of building a business process map is to enable us to see how the organization translates its customers' requirements into a product or service – the basic process of any organization and at the root of the definition of quality. This map then enables us to break down the major business processes into more manageable sub-processes and tackle improvement on the whole process rather than just the bits for which our functions are nominally responsible. An example is shown in Figure 7.4

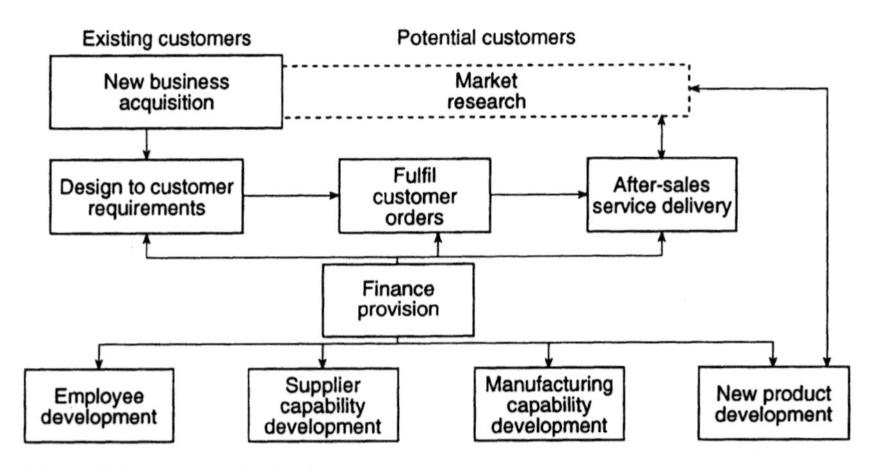

Figure 7.4 Example of a business process map.

Start by viewing your business as a process with customers and the market at one end and supplied assets (material, human and financial) at the other. The activities which convert the latter into the former sit in the middle as shown in Figure 7.5.

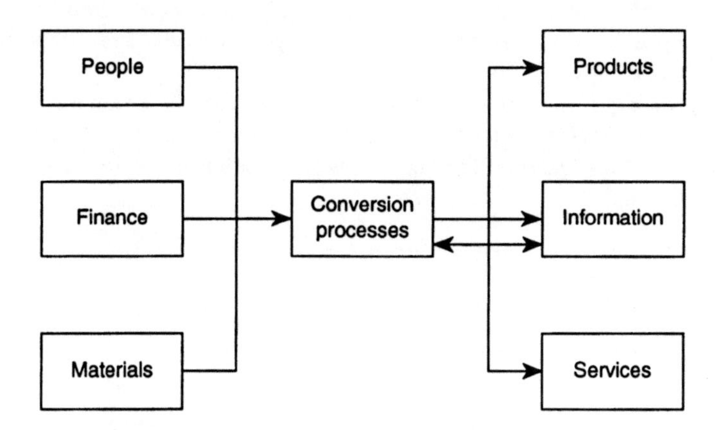

Figure 7.5 A business as a process.

Now list the **processes** involved in making this transition. As a guide, a process description should contain a verb and an object so that the statement says what is being done to achieve what end. Examples of good and bad process descriptions are given below.

Bad

- Financial accounts
- Personnel
- Training
- Procurement
- Production
- Marketing

Good

- Provide management information
- Resource employees
- Develop employees
- Provide materials
- Fulfil customer orders
- Import customer requirements

As a general rule, aim for no more than ten processes. Remember that the scope of the process is wide. Employee development, for example, takes the process from the point at which the individual makes first contact with your organization through to their exit.

2. Define the **key** *processes which have major implications for the organization's critical success factors*

This can be achieved by borrowing a tool from marketing – the importance vs. satisfaction map, sometimes used in establishing customer requirements from market research. The method involves examining each business process and rating it for satisfaction. If it performs well it gets a high score – you can decide your own scale but the results form one axis of the chart on which the results of the analysis will be placed. The other axis represents the importance of the process, i.e. the degree to which it impacts on each critical success factor. Again a scale should be used and the score for each CSF added together to produce an overall importance or impact rating.

Each process is now placed on the chart at the point at which the satisfaction score intersects with the importance score. Action is based on an examination of the area of the chart in which each process falls, as shown in Figure 7.6.

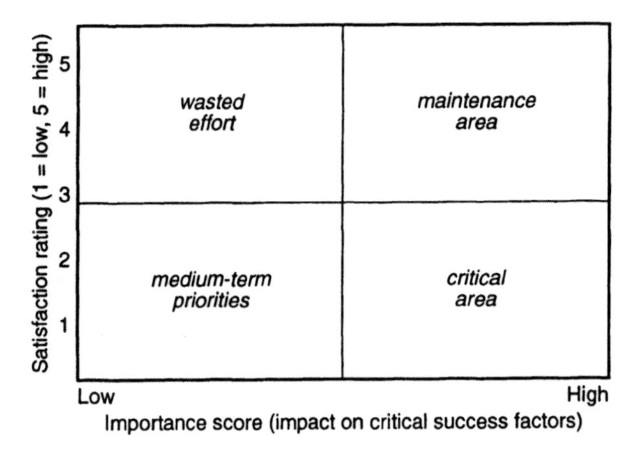

Figure 7.6 Business importance vs. satisfaction map.

The critical area represents the area for attention in the context of re-engineering. The maintenance area requires monitoring and the application of continuous improvement processes. The medium-term area and wasted effort areas are self-explanatory. This map should be revisited at least annually.

3. Establish teams to review the process

The teams should consist of the key stakeholders in the process and a process owner who should be one of the top management team. The process owner may not attend all meetings but should act as a mentor to the re-engineering team. The team should also have access to an IT specialist, ideally as a member of the team itself.

Team members need to be trained in process analysis techniques and methods of creative and lateral thinking.

4. Map the process as it is now

The techniques for this were explained earlier in this chapter. It is a good idea to calculate the cost of quality inherent in the process as the flowchart of the process is put together.

5. Go back to basics and describe the process as it could be if this were a greenfield operation

This requires input from the IT specialist and the use of the techniques of lateral thinking. The team must be able to challenge very basic tenets or paradigms about how a process has to be organized. It is a good idea to start out by listing all the assumptions inherent in the process and to consciously challenge them one by one. For example, a few possible assumptions inherent in the process of employee development are shown below – you can probably think of others:

- The company has to have employees
- Development is a management function
- Development is for promotion
- Development has to be job-related
- Development costs money
- Potential employees need to be interviewed
- Technical staff must have qualifications

The value of the IT specialist is in bringing knowledge to the team of the potential which information technology can bring to the process. Some examples of IT support which can significantly quicken, improve and shorten processes are:

- EDI – electronic data interchange
- DIP– document image processing
- Teleconferencing
- Networked software
- Portable computers, faxes and phones
- Modems and home-based operating
- Bar-coding
- Shared databases with customers, suppliers, etc.
- So-called **expert** systems
- Robotics

Drawing up the new ideal process involves both streamlining and mistake-proofing (they often go hand in hand) the process. This means critically reassessing the process diagram for duplication, check points, loops and by-products, instituting automated processes for human routines, improving reliability at the front end and examining the true value of the process. Paperwork should be assessed for simplicity, necessity and clarity.

Value-added can be determined by asking three simple questions:

1. Is it necessary to produce the end result or output of the process?
2. Does it contribute to customer satisfaction?
3. Does it contribute to other essential business functions?

If the answer to all the above is no, then the process is adding no value and should be eliminated. Finally a new process map is drawn up based on the ideal process.

6. Adjust the process to take account of unavoidable internal or external requirements

In the real world we still have state bureaucracy, corporate rules and other undesirable but unavoidable requirements placed on our processes. The team will soon become disillusioned if it is allowed to put up scenarios which are immediately knocked down by the organization. They must therefore adjust their ideal process to take account of the constraints in which it **must** operate. This process, however, should be rigidly applied to see whether the constraints are real or assumed.

7. Implement the new process

The new process will probably have several characteristics which make implementation difficult and therefore something which needs to be carefully planned and managed. For instance:

- The work may involve significantly fewer people, giving rise to potential redundancy.
- Decision-making may be devolved to the natural work group giving rise to loss of control by management and again potential redundancy.
- Work may be moved requiring relocation of people, offices and equipment.
- New technology may be brought in requiring project management.
- Checks and controls may be reduced creating customer concerns.
- Human tasks may become more complex and require greater decision-making and analytical skills as opposed to manual processing skills, and requiring education or workforce changes.
- Opportunities to identify and rework errors may disappear giving rise to higher external failure costs.
- Work may be outsourced requiring new forms of control (and possible transfers of staff under the business transfer regulations in the UK).
- Flexibility and multi-skilling will require educational support and possible career changes.

7.2 PROBLEM IDENTIFICATION

> There are hundreds of easy ways to solve difficult problems – the only thing they have in common is that they are all wrong
>
> The only generalisation you can rely on is that all generalisations will lead you to the wrong conclusions and will get a lot of people cheesed off in the process
>
> Richard Barnes

Problems arise through some variation from the planned or expected outcome. Failure to understand the nature of this variation, as explained under Process control gives rise to incorrect identification of problems. The single biggest mistake made in problem solving is in identifying the solution before identifying the cause, let alone the problem itself. Ready, aim, fire! becomes ready, fire, aim! or worse, fire, ready, aim! On our courses one trick employed is to hand round some photographs which are out of focus and ask people what they think the problem is. Occasionally someone will reply flippantly (but correctly) that they are out of focus. Equally flippantly someone will ask (again correctly) whether the photograph was meant to be of an exquisitely defined twig instead of the fuzzy gorillas in the background.

Most people however, feeling an affinity with the poor old tutor, will shout out a range of ideas ranging from camera shake to auto focus failure. The exercise ably demonstrates a Western tendency, perhaps springing from our entrepreneurial culture, to fly by the seat of our pants and take educated (or should that be experienced?) guesses rather than find out the facts.

One large manufacturing company had complaints from neighbours in the local village about noise from the extractor fans during the night shift. Investigations by the engineer's staff confirmed that the fans were noisier than usual and it was concluded that this was natural deterioration in the fan housings. The company spent £11 000 on noise reduction equipment to surround the fan housings. Two weeks later, a small pump in the basement gave out and the noise from the fans suddenly dropped off. The pump, which was located next to the outlet ducting, had been oscillating for a number of weeks sending vibrations through the ducting which emerged as excess noise from the fans. A few hours investigation and a £50 replacement pump would have saved the company over £10 000.

What the engineers had done of course was to fall into the trap of jumping from step A to step E in Figure 7.7

The other face of this tendency to jump to conclusions is our love of generalizations or stereotyping. Life is a lot simpler if we can put it into neat little packages – it avoids us having to explore the situation in any detail. Jack is *always* late, Sheila doesn't give a damn about *anything*, management have *no* commitment, people *never* take enough care, are all examples of this tendency. The following statements were taken from real people defining real problems. Spot the generalizations:

The order clerks in the Commercial Office continually make mistakes on the order forms that come down to us.

Our advertising agents are always getting errors into the job advertisements, no matter how careful we are in drafting them.

Nobody ever replies to my memos asking for information.

Our stationery suppliers are unreliable.

Getting the initial definition of the problem right is the most important and least well managed of all the stages of problem solving. A good problem definition should be precise, contain as much factual data as possible and must not assume any causative factor. It is very important that a formal statement of the problem is produced so that these criteria can be checked off. This should happen irrespective of whether the problem is being handled by a group or an individual. Suitable forms to aid problem definition should be developed by the steering group member responsible for corrective action.

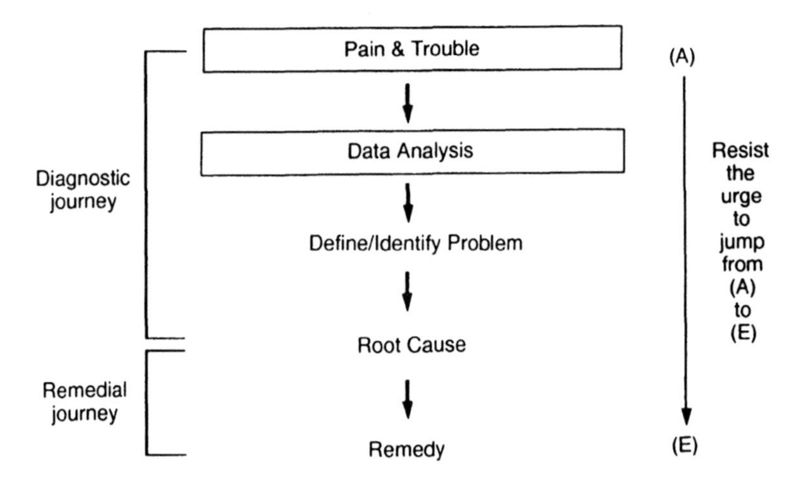

Figure 7.7 Stages in problem solving.

Identifying what problems to address is the second most important issue. We are using the word problem in this section in a wide sense as not just a departure from an accepted standard but also undiscovered gaps between desired and actual performance. Sometimes we don't know we have a problem; the ubiquitous **acceptable quality level** or economic level of quality is a major source of undiscovered problems. They are also disasters waiting to happen. One supplier of photographic paper for a major multinational had an accepted quality standard in relation to holes per thousand metres of paper. This standard had been agreed with the customer and there was no pressure from the customer's side to improve it. In Switzerland, a competitor upgraded his equipment and offered the customer a significantly higher standard as a result of better process control technology. Overnight almost a third of the supplier's business was lost to his competitor.

What this story illustrates is the fallacy of acceptable or economic levels of quality. It is what Phil Crosby really means by zero defects, as opposed to how Deming mischievously chooses to interpret it. The reality is reflected in the graph shown in Figure 7.8.

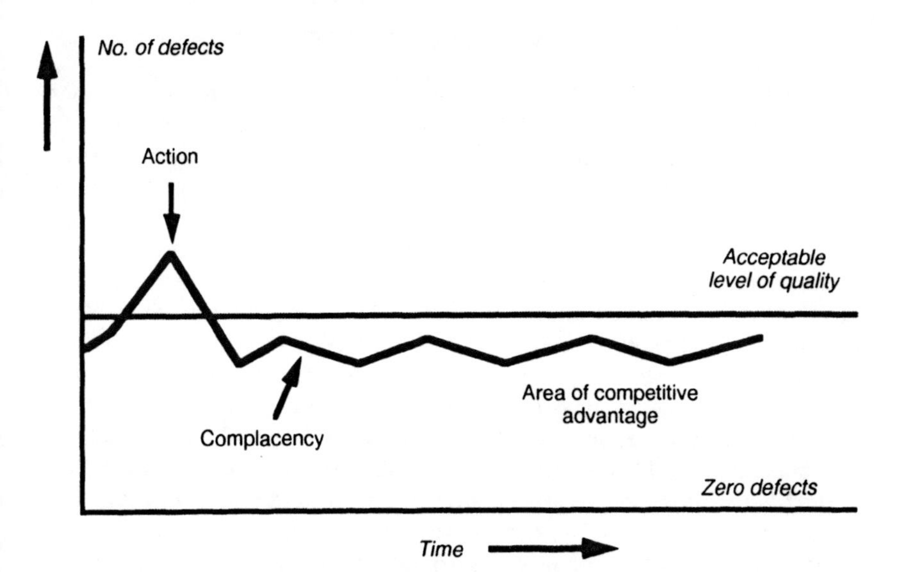

Figure 7.8 The implications of the *zero defects* concept.

The first step then in problem identification must be to make sure you have your sights on the right target. There are several techniques for problem identification utilized in TQM. They can roughly be divided into two types; those which identify failures to meet existing requirements and those which identify undiscovered gaps and thus set new targets:

Systems for identifying existing gaps

Quality system audits (a requirement of ISO9000)
Quality Assurance third party corrective action notes
Customer Complaint analysis

Systems for identifying undiscovered gaps (i.e. new targets)

Cost of Quality studies
Benchmarking

Formal problem reporting or error–cause removal systems
Departmental Purpose Analysis (see pages 203–206)
Customer and employee surveys (see pages 88–93 and 169–174)

We have already dealt elsewhere with customer and employee survey and complaint data and so we will concentrate in this section on the more pro-active systems of cost of quality, problem reporting and benchmarking.

The systems to identify existing gaps are more reactive in nature, but are nevertheless useful sources of quality problem knowledge.

7.3 COST OF QUALITY – AN OVERVIEW

Chapter 4 briefly outlined the definitions of quality costs as a means of demonstrating the need for change and gaining top management attention to the potential gains from quality improvement. Below we deal with the development of quality costing as an ongoing mechanism to monitor performance.

Put simply, quality costing is the means of monitoring the causes and effects of doing things wrong. It is a powerful technique for getting management attention, for bringing 'quality' into the boardroom, for prioritizing problems and for showing improvements. The best way to measure Total Quality is to calculate the real effect on the business of failing to achieve requirements. When a requirement is not achieved correctly the first time, the full cost impact is classified as a cost of non-conformance (CONC).

Non-conformances are thus measured in £s rather than in numbers, percentages, etc. This focuses attention, enables prioritization and leads to the justification of corrective action. It also provides us with a common denominator of cash with which to talk about quality rather than having to deal with an often vague and disparate set of quality symptoms.

The experience of many companies has been that focusing on the most urgent improvement areas is facilitated readily by the employment of Quality Cost Measurement. These companies and many others have shown again that, as the photograph paper example demonstrated, there is no such thing as an economic quality level, and that it is always cheaper to get it right first time. Of course, finding out what is right is part of the fun.

Two out of three main quality cost elements are monitored:

7.3.1 Cost of conformance (COC)

This can be defined as the cost of investing to ensure activities are carried out correctly the first time and every time and problems are prevented. Examples are training, selection testing, inspection and process control. These can be further subdivided into **prevention** costs, which are concerned with not allowing problems to occur in the first place by foolproofing the system, and **appraisal** costs which are concerned with checking for them after they have occurred.

7.3.2 Cost of non-conformance (CONC)

This is the cost incurred by failing to achieve an activity correctly. Examples include scrap, downtime, errors, rework and lost business. These failure costs can be split into **internal** and **external** failures, depending on whether the customer experiences the results of the failure or whether the problem is contained within the company.

These two can be referred to as the **Total Quality Cost**. In most organizations 80% of the Total Quality Cost is accounted for by the cost of non-conformance.

7.3.3 Essential costs

These represent the essential costs wholly attributable and necessary for the product or service to be as specified, such as the minimum necessary material, labour, energy and information. These are considered unchangeable and so are not formally recorded.

7.3.4 Magnitude of total quality cost

Studies and the experience of many companies have shown that a typical manufacturing organization wastes between 10% and 25% of its turnover. In some service industries figures as high as 40% have been cited. This is not all that surprising when you think that manufacturing has 100 years of quality control experience behind it whilst the white collar areas of a business, service industries and the professions have hardly any. Well over half our quality problems are created in the staff areas yet 90% or more of our quality controls are in manufacturing and on blue collar operators.

Such expenditures add nothing to the value of the product or service. Reduction of the costs of non-conformance can also lead to substantial reductions in conformance costs as well. This is due to the reduced need to check, test or inspect. Rank Xerox's manufacturing operation at Mitcheldean for example, was able to reduce its inspection of incoming parts from 80% to 15% (primarily new parts or suppliers) whilst reducing reject rates on supplied parts from 30 000 per million to 300 per million (the target is 100 per million).

A major factor in making large savings is the identification of multiplier effects. For example, a mistake or omission in the specification of a major product, such as an air traffic control system, may only cost some tens of £s to rectify before design work commences. However, if the problem is not recognized until the commissioning stage, the cost of rectification will have multiplied by many hundreds, if not thousands, of times. The quality spiral giving rise to this multiplier effect is illustrated in Figure 7.9.

This multiplier effect, sometimes known as **process value-added**, is also apparent in management processes, such as those which control production

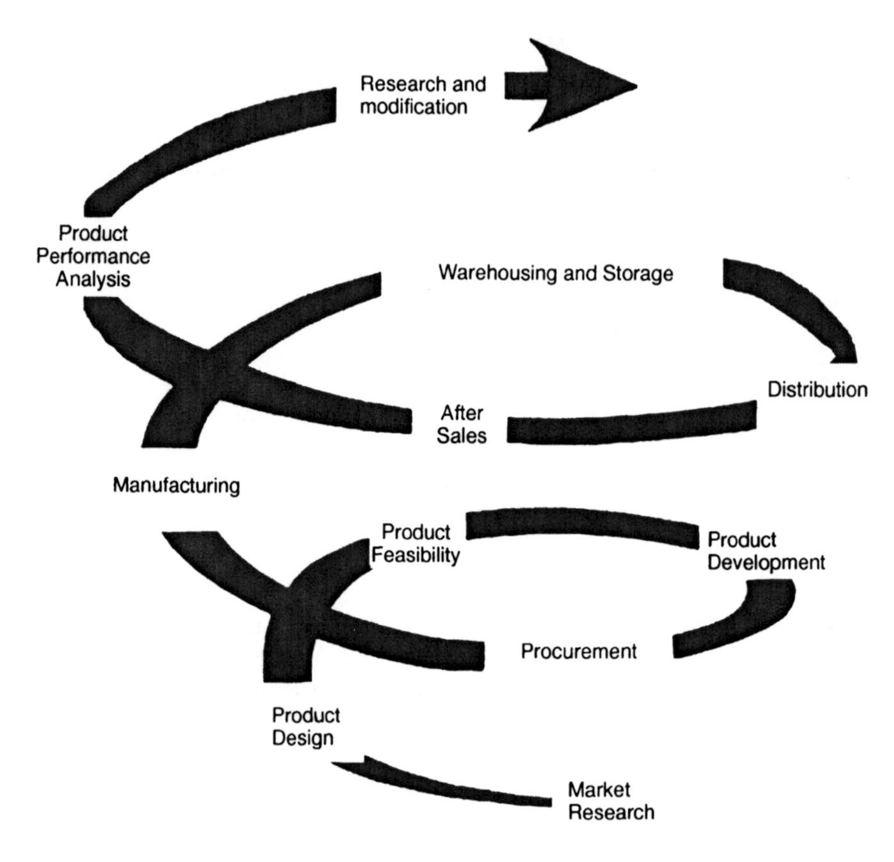

Figure 7.9 Quality spiral showing how process value is added

operation in a manufacturing company, or the activities of people in a service company. For example, a mistake made in entering an incorrect part number into a computerized manufacturing requirements system could have a prevention cost, at this stage, of £5. After processing the rectification cost could be £100. If the mistake is not discovered until after production has been completed, then the true cost of non-conformance could include scrap, rework and investigation costs as well as the replacement cost. The cost will, by this time, be in £1000s. Many companies have found that such management system mistakes account for up to 70% of their 'shop floor' costs of non-conformance. The study by Philips referred to earlier graphically illustrates the problem.

At first sight, figures of 20 to 30% of turnover for a company's Costs of Quality seem exaggerated to many managers. As the following examples illustrate, these figures are not just opinion, they are fact. For instance, when starting out on TQM, STC Telecommunications found they had, as a percentage of sales, a Cost of Quality of 24% in their stable products, 31% in stable exports and 41% in new products. British Telecom originally estimated their Cost of Quality at 25% of turnover – representing 2½ to 3 billion pounds a year. At NatWest Bank it was estimated that 25% of its operating costs were taken up in non-added-value activities resulting from jobs not being done right the first time. BAe Dynamics found that their manufacturing quality costs were 11% of the cost of production, with almost half of those costs being failure cost and less than a quarter in prevention. Over several years concentrated effort at the process of quality improvement, companies report impressive results:

- Bridgeport Machines Ltd halved their quality costs and doubled their sales turnover between 1982 and 1990
- Standfast Dyers and Printers of Lancaster report a reduction in their cost of quality from 20% to just 7% of sales value in 5 years
- ICL's Kidsgrove plant reduced its Cost of Quality by 10%, saving £9 million in 2 years.
- Philips Components of Blackburn report elimination of 60% of quality costs over 6 years of TQM
- Allen Bradley (part of Rockwell International) reckon to have saved 82 million dollars through TQM. They lowered the cost of quality on programmable logic circuits from 13% of sales to less than 5%. They estimate that TQM has saved £13 for every £1 invested.
- In one of Ciba Geigy's main operations, they reduced the cost of repeated lab work from 60 000 dollars per annum to 25 000 through the introduction of better forms and procedures to improve the identification of internal and external customer requirements.
- Philips of Germany introduced TQM during the early 1980s. By 1989, they had reduced their cost of quality as a percentage of production value from 24% to 11%. Failure rates in production were reduced by 20%.
- Baxter Healthcare (USA) report a reduction of 2.9 million dollars in the cost of quality in less than 3 years under TQM. This is valued at 3% of the annual value of production added at the plant.
- Rhone Poulenc Fibres reduced the cost of quality from 20% to 16% early in the TQM process. They believe they have had a five-fold return on an investment of FrF10 million.
- Milliken claim to have achieved a 60% reduction in the cost of nonconformance since 1981.

7.3.5 Quality costing implementation

The method of implementing a quality cost study will differ for individual companies depending on the purpose for which the study is intended, the depth of resource it is prepared to commit and the maturity of the financial and other data collection systems in existence. However, there are some commonly accepted ways forward which recognize that the establishment of a quality costing system is a gradual process, moving from a very rough guide to action in its early stages to a more sophisticated and comprehensive database and early warning system in its maturity. Establishing such a system can take 5 to 10 years.

The steps involved will contain most, if not all, of the following issues and decisions:

1. Appoint a person on the steering committee responsible for seeing the process through. The company Finance Director often adopts this role.
2. Establish the objectives and operating protocols for the system at the steering committee; these should include:
   - What the purpose of the system is; e.g. to identify improvement projects or to build a comprehensive financial system for ongoing cost collection.
   - What categories of cost will be used; e.g. COC/CONC, PAF (prevention, appraisal, failure), etc. A common language must be established at the outset.
   - What criteria will be used for including costs. Decisions here will depend on the objectives of the system. If the objective is to identify improvement targets, there will be little point going to great lengths to collect data about appraisal and prevention costs except as a means of demonstrating the payback from such activities. Nor will it be of much value at this stage to collect costs for items which are considered unchangeable. A figure should also be set as a minimum below which costs can be ignored, say £500.
   - How data will be verified. It is usually the role of the finance department to guard against double counting and over-enthusiasm by the quality department and to ensure that the data collection systems are soundly established and have credibility amongst users.
3. Establish a Cost of Quality team. They will be responsible for collecting the data at source and should therefore be drawn from across all departments of the company.
4. Train the Cost of Quality team in the methods and approaches to quality costing.
5. Conduct an initial pilot study of quality costs in a specific area of the company's operations and report these to the steering committee.
6. The steering committee will need to make a number of decisions based on the report from the Cost of Quality team. These will include:
   - How and where the costs will be recorded. This may include the establishment of a memorandum account for quality costs
   - The coding to be used for the types of cost uncovered and how

these will fit into the other financial systems, e.g. cost centre codes
- Which costs are to be collected on a company wide basis and what category they will fall under
- The establishment of a **bellwether** or **fast-track** project aimed at reducing an important and visible element of non-conformance costs
- Reporting and display mechanisms
- Steps to be taken to expand the system to include any missed functions or categories of cost. Decisions may also be required on how to improve the accuracy or focus of the cost data; e.g. splitting COC into prevention and appraisal.

The respective roles of the various sub-groups involved in quality costing can be summarized as shown in Figure 7.10.

7.3.6 Sources of quality cost data

One of the first tasks of a team charged with establishing a quality cost study is that of locating sources of information on quality costs. Unfortunately, many quality costs, particularly the costs of non-conformance are not recorded anywhere. Outside the production area, the information is even more difficult to find, yet it is in these administrative areas that the greatest potential costs occur, precisely because they have never been subject to quality control or costing.

Sources of cost data should follow a top to bottom and easy to difficult route. In other words, start with the major accounts codes and reports and work down to departmental and individual levels. At the same time start with secondary data (i.e. produced for other purposes, such as the payroll) and work down to primary data (i.e. specifically recorded for the quality costing system). The aim is to establish the maximum amount of data with the minimum amount of effort. New, primary data, cost collection systems should only be established if they are to be used to improve some aspect of organizational performance; this is not an academic exercise!

Sources of secondary data to consider are:

Payroll costs – For those people directly attributable to quality costs. For example COC costs might include the training manager, quality department staff, inspectors, supervision. CONC might include staff purely reserved for handling problems, rework or customer complaints, overtime and other unplanned excess payments. In most cases, a split between COC and CONC will need to be made.

Overheads – A proportion of building and administrative costs needs to be added to the payroll costs

Capital budget – Costs of specialist equipment such as test or calibration systems

COQ STUDY – ALLOCATION OF RESPONSIBILITIES

ACCOUNTS DEPT	QUALITY DEPT	PILOT TEAM
CONTROL	*FACILITATE*	*DO*
Establish Memorandum account	Facilitate use of data collection and extraction tools by staff	Data location and extraction
Establish approved codes	Assist teams with process cost modelling	Problem reporting
Approve forms and report designs	Allocate codes to agreed costs from approved list	Allocate responsibilities for cost collection and reporting
Verify costs for rigidity	Determine COC or CONC categories	Recommendations on codes and forms/reports layout
Check for double counting	Forms and report design	Recommendations for primary data collection systems to be set up
Agree overhead attribution	Arbitrate on departmental allocation of costs	Process cost modelling
Produce summary report for Board and Steering Committee	Cause investigation in key areas	Process mapping
Determine key ratios and their interpretation	Pareto analysis	
	Recommend improvement initiatives to the Steering Committee	

Figure 7.10 Typical responsibilities in a quality cost study

Purchase ledger – Costs of purchases such as materials for testing, samples, external quality services, audits. Also costs of holding safety stock purchased to allow for emergencies

Departmental accounts – Headings attributable to conformance or non-conformance, such as waste, premium freight charges, demurrage, outstanding bills, warranty costs, etc.

Variance reports – Reports from budgets, projects or departments showing departures from plan as a result of quality failures

Manufacturing reports – expenses, scrap and downtime analyses

Travel expenses claims – these can sometimes reveal clues to follow up

Maintenance reports – Engineers' reports

Time sheets and job cards

Inspection and test reports

Audit reports

Process descriptions – Using the process cost modelling method, reviewing process flowcharts and descriptions for points of rework, prevention and inspection costs and highlighting these.

Where payment is received for scrap, or by the customer for repair/ rectification work, these costs should be entered as income. Some experts recommend that income from scrap is not deducted from wastage costs, but this can result in a kind of double counting. The income however should be itemized separately.

7.3.7 Methods of extracting quality cost data

There are five main methods of extracting data for a Quality Costing System. These are listed below in descending order of both exactitude and ease of collection. Methods 4 and 5 are usually reserved for specially agreed costs which are to be subject to improvement activity.

1. **Account head analysis** – existing information from monthly results or departmental analyses. Examples are given above.
2. **Payroll analysis** – also from existing information as explained above. Note that in this item and the preceding one, some headings may include both 'error-free costs' and quality costs. In such cases it may be necessary to refer back to source documents to allocate the costs. It is important that where positions are attributed to quality, they carry the full overhead cost of the salary to reflect costs which would not otherwise be incurred such as their share of space, equipment,

training, health, pensions, welfare, etc.

3. **Work reporting** – these data may be either secondary (i.e. from existing information as in items 1 and 2) or primary data specifically collected for the purpose of costing. Information may be obtained from such instruments as time cards, job sheets, etc. Examples of work reporting costs might be engineers' time spent returning to stores or waiting for a part, time spent on a litigation claim, salesman time spent on a complaint amelioration, time spent debugging a computer programme.
4. **Error costing** – multiplying the averaged cost of one type of non-conformance by its frequency to produce a total cost. This technique is especially useful where a number of different people are involved and an error has effects in more than one area, e.g. a design change, an order processing error or returned goods. It is the only realistic way to deal with frequently occurring low-cost problems without generating a great deal of administrative work. The principle is rather like a Local Authority Maintenance department's schedule of rates. It doesn't try to cost every tap a plumber repairs, but establishes a standard average rate for the time and materials used in a normal call, even though the actual time spent may vary widely on individual jobs.
5. **Surveys** – although the least exact method, they are the only available way of quantifying many managerial jobs. Individuals are continually moving from development type jobs to error correction, review or just spending excess time on items, e.g. meetings going on too long. Time management analyses taken at suitable intervals are a useful way of measuring progress and identifying problem areas for attention. This could be in the form of a periodic log over a week showing what is being done each 15 minutes during a day.

7.3.8 Reporting of quality costs

There are probably as many different reporting formats for quality costs as there are systems for collecting them. There is only one golden rule when reporting costs, which most accountants will be familiar with:-

WHATEVER CATEGORY YOU PUT A PARTICULAR COST INTO, REPORT IT IN THAT CATEGORY AT ALL TIMES. CONSISTENCY IS MORE IMPORTANT THAN ARGUMENTS OVER WHETHER IT IS A PREVENTION OR APPRAISAL COST.

Several types of report are required for interpretation and use of the data.

The main **customers** for the data are:

The steering committee – for assessing progress, identifying projects and evaluating investments in prevention activity

The board or executive team – for assessing the value of the activity and its relationship to business costs and results. Use of business ratios as suggested below are helpful

Department heads and local or cross-functional Quality Improvement Groups – for identifying local improvement projects and recognizing contributions from employees.

There is only one purpose for the reports produced. **TO STIMULATE IMPROVEMENT ACTIVITY** – either at the management level in changing or improving the systems which cause quality costs, or at the local level in targeting areas for reduction of defects or increases in right first time performance.

Where companies operate more than one facility, it is important that common reporting styles, coding and categories are used so that comparisons can be made between them by the board. Some practitioners, notably Juran, recommend use of a scoreboard to stimulate competition between plants and departments. Such tools should be used with caution however as they may invite cries of **foul** if common causes present in certain areas are not taken into account in reporting the results.

7.3.9 Business ratios

Several ratios are of value in linking quality cost data to key business variables. Among the common ones, often shown as percentages, are:

$$\frac{\text{Cost of Quality}}{\text{Sales value}} \times 100$$

$$\frac{\text{Project Cost of Quality}}{\text{Project value}} \times 100$$

$$\frac{\text{Cost of Quality}}{\text{Value added}} \times 100$$

$$\frac{\text{Cost of Quality}}{\text{Production costs}} \times 100$$

$$\frac{\text{Cost of Quality}}{\text{Product unit cost}} \times 100$$

$$\frac{\text{Cost of Quality}}{\text{Payroll costs}} \times 100$$

The ratio used should be viewed in relation to change in other factors which may affect them; automation, redundancies, change in product portfolio make-up, etc., which can influence the results dramatically.

The system should be reviewed on a regular basis and improvements to be made identified.

7.4 BENCHMARKING

Benchmarking is another example of a word which has been reworked for use in Total Quality (just as the word quality has). This trend toward reinterpretation of existing words and invention of new non-words is one of the more irritating aspects of the whole quality movement, but when you have so many consultants like us involved in it, it becomes rather inevitable as we all compete to sell our models.

Traditionally we think of a benchmark as a particular standard or typical example. In TQM it refers to a **best-in-class** standard for a particular product, service or process. Best-in-class is also given a wide definition; it is taken to be a universal examination of best practice or performance and not simply confined to the industry sector concerned.

For example, Rank Xerox use a system of benchmarking across industry sectors to locate the best at a particular process: e.g. production schedule compliance is marked against Cummins Engines and invoicing against American Express. Xerox benchmarks its distribution capability against L L Bean.

ICL have benchmarked their distribution system against Marks and Spencer's. Motorola chose Domino's Pizza and Federal Express to study delivery process speed. Apple and Compaq have even collaborated and benchmarked certain processes against one another.

Benchmarking is not always confined to traditional manufacturing or service areas; it can also be turned on internal service standards. For instance, IBM, Motorola, AT&T and Xerox have all benchmarked their training programmes against recognized leaders in that field. Figure 7.11 features the Rank Xerox approach to benchmarking which helped them to win the first European Quality Award in 1992.

Many of our leading-edge businesses have made substantial investment in this process. Johnson and Johnson have benchmarking teams throughout Europe and Asia whilst Digital Equipment Corporation has an electronic benchmarking network that covers the globe, linking thousands of people.

Benchmarking, like the cost of quality exercise, is not a short-term initiative. It should be viewed as a visionary process for continually raising the sights of all functions in the organizations. It is zero defects, and then some!

Typically the person charged with measurement on the steering group should take overall responsibility for the establishment of plans for the instigation of the system, but all areas of the organization need to be involved in defining the targets and the data sources. A separate sub-group of the steering group will be required in most larger outfits.

7.5 PROBLEM REPORTING SYSTEMS

Those organizations which have established a formal quality management system conforming to the ISO9000 series of standards will be familiar with the system for identifying corrective actions. The problem reporting system (sometimes known as error-cause removal or ECR) simply extends the origination of corrective action requests to all areas of operation to allow any employee at any level to have an input into a formal problem identification, review and removal process. In the way it operates, it does not differ greatly to the way many organizations run their suggestion schemes; the difference is that employees do not have to have an idea, only a problem in need of a solution. The objectives of the problem reporting system are:

1. To support the error-friendly environment by making it OK to admit to problems, knowing they will be tackled rationally and without blame.
2. To extend the franchise for getting action initiated to every department and employee.
3. To establish a common approach to problem elimination and assist in prioritizing action.
4. To encourage a team approach to cross-functional problems.
5. To assist in maximizing employee involvement.

7.5.1 Organizing the system

The responsibility for this element of the plan lies firmly with the person tasked with establishing the corrective action system. However, particularly in the early stages of its use, the task holder may be swamped with action requests and will require assistance. Again the use of a specific sub-group of the steering group, acting as a review panel should be established. This team must be trained to understand the principles of statistical variation and be able to interpret when a problem being referred to them is likely to be the result of a special cause or a cause common to the system. This will determine whether further study, using statistical control techniques is required. One of the faults of most corrective action systems is that they treat the majority of problems as though they each had assignable causes and thus do nothing to reduce the overall frequency of problems in the system.

Benchmarking was used in the development of the Leadership Through Quality strategy, and extended far beyond competitive analysis to include the identification and study of the best in class for all areas of process management relevant to the business. Benchmarking is defined by Rank Xerox as 'the continuous process of measuring our products, services and processes against our strongest competitors, or those companies renowned as the leaders'. Benchmarking is used to identify what the company needs to do to become the benchmark and therefore the goals Rank Xerox needs to set itself. The benchmarking process, described in full by Robert Camp in his seminal work on the subject has ten steps, in five phases, as shown below.

A example of benchmarking in the company's European Manufacturing Organization is shown in the following page (Figure 2).

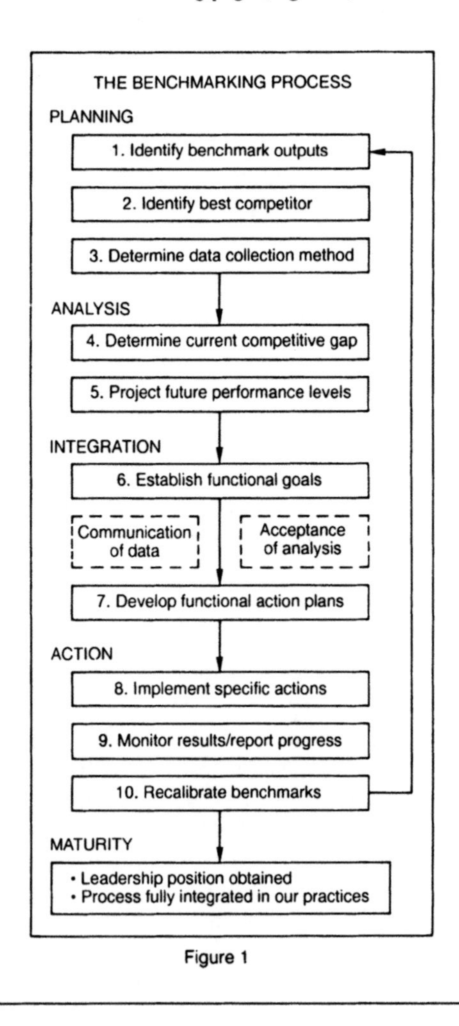

Figure 1

Figure 7.11 Rank Xerox's Benchmarking process.

Over the last eight years the company has conducted a wide array of benchmarking programmes in all aspects of the business (Figure 3). During this time the use of benchmarking broadened from being a function specific activity in individual operating units to becoming increasingly cross-functional and international. The strategic focus of benchmarking has been enhanced by integrating it into the company's planning and policy deployment process. In 1990 Rank Xerox set its 3 to 5 year planning goals at benchmark level. These goals in turn were translated into annual objectives.

Individual operating units use benchmarking locally as a key enabler to achieve their own 'vital few' objectives, while headquarters coordinates pan-European benchmarking programmes in support of common goals across the company.

Benchmarking started off as a tool to identify industry cost and performance standards and to set goals. The focus evolved to cover world-class leaders in any industry and to emphasize the process used as well as the targets achieved. Benchmarking became a mechanism to provide insights into how these cost and performance standards could be achieved or exceeded, via the ownership of these findings and development of appropriate internal action plans.

In this respect benchmarking across the company is viewed as a learning experience both for the people involved and the company as a whole. It is used to help define the productivity improvements needed to compete effectively and act as a key catalyst for change.

The experience with benchmarking in Rank Xerox suggests the following keys to successful implementation. Firstly as with any venture of organizational significance it is critical to have leadership and commitment at the top. Equally important is to have appropriate discipline associated with selective flexibility and adequate training in implementation of benchmarking projects. Finally it is critical that benchmarking is integrated into the management process of an organization. Benchmarking can become a resource intensive activity and should not stand alone from a Total Quality strategy.

BENCHMARKING IN ACTION AT RANK XEROX

Purchasing Commodity Teams used Benchmarking with a major competitor to improve the way the company designed, procured and manufactured sheet metal products.

This study revealed key enablers to help improve the strategy to drive commonality in material (a reduction from seven types and 13 thicknesses in 1986, to two types and eight thicknesses by 1989 with a 11.5% cost reduction), leading to a review of design practices to reduce processing steps, and an improved approach to acquiring the tooling.

Information gained from such a benchmarking exercise is shared with the supplier base as part of a partnership approach to Materials Management.

Figure 2

EXAMPLES OF BENCHMARKING PROJECTS

- Price Level in Europe
- International Account Mgt
- Product Delivery (Order Mgt)
- Data Centre Study
- Compensation and Benefits
- Configuration Management

Figure 3

Figure 7.11 – contd Rank Xerox's Benchmarking process.

The role of this review panel is not to attempt to solve the problems referred to it but to instigate the correct type of investigation and direct the problem into a suitable arena for its proper execution. As such the group acts principally as an interpreter and clarifier of the problem and then as a post office to re-direct it.

An outline of a suitable administrative process for a corrective action request review panel is shown in Figure 7.12. Note that at various stages the question has to be asked as to whether the issue is likely to be a system problem; i.e. the responsibility of management or a control problem – the responsibility of employees.

7.5.2 Prioritizing problems

The outputs from the systems set out on pages 215–216 will be many and various and one of the greatest difficulties for those embarking on TQM is in deciding on the priorities for action.

On the face of it, an exhortation by us to do those things first which bring the greatest benefit may sound rather obvious and simplistic. However, we have found that most organizations and individuals find it easier to do the simple things first and have difficulty in getting round to the effort required to penetrate those areas which are the real gold mines in TQM. Instead they tinker around with sandwiches in the canteen or photocopier waste, often with the excuse that they are just starting to get a feel for the process. In fact they rarely get past this stage and in targeting these kind of trivia, actually do damage to the whole process by downgrading its importance and potential.

The Pareto principle (so called after Wilfredo Pareto, a nineteenth century Italian economist who made the observation that in industrialized nations, 80% of the population owned 20% of the wealth and that this ratio was apparent in many other areas of life), known as the 80/20 rule, it is a valuable tool for sorting the wheat from the chaff in problem solving and improvement activity. Or, as Juran is fond of saying, sorting the vital few from the trivial many. 80% of the quality cost savings in TQM will come from 20% of the activity. 80% of new business will result from 20% of the improvement initiatives.

Because there are so many potential activities under the TQM umbrella, deciding which ones to address can easily turn into a lottery. The Pareto method is a simple but effective tool to prioritize improvement by taking the costs, frequencies, potential benefits, logical rankings of ideas, etc. and stacking these on a graph in order of magnitude (see example in Figure 7.13). The graph is then used to select the most beneficial projects which will bring the greatest reward in the targeted criteria – cost reduction, morale improvement, complaint minimization, new business, etc.

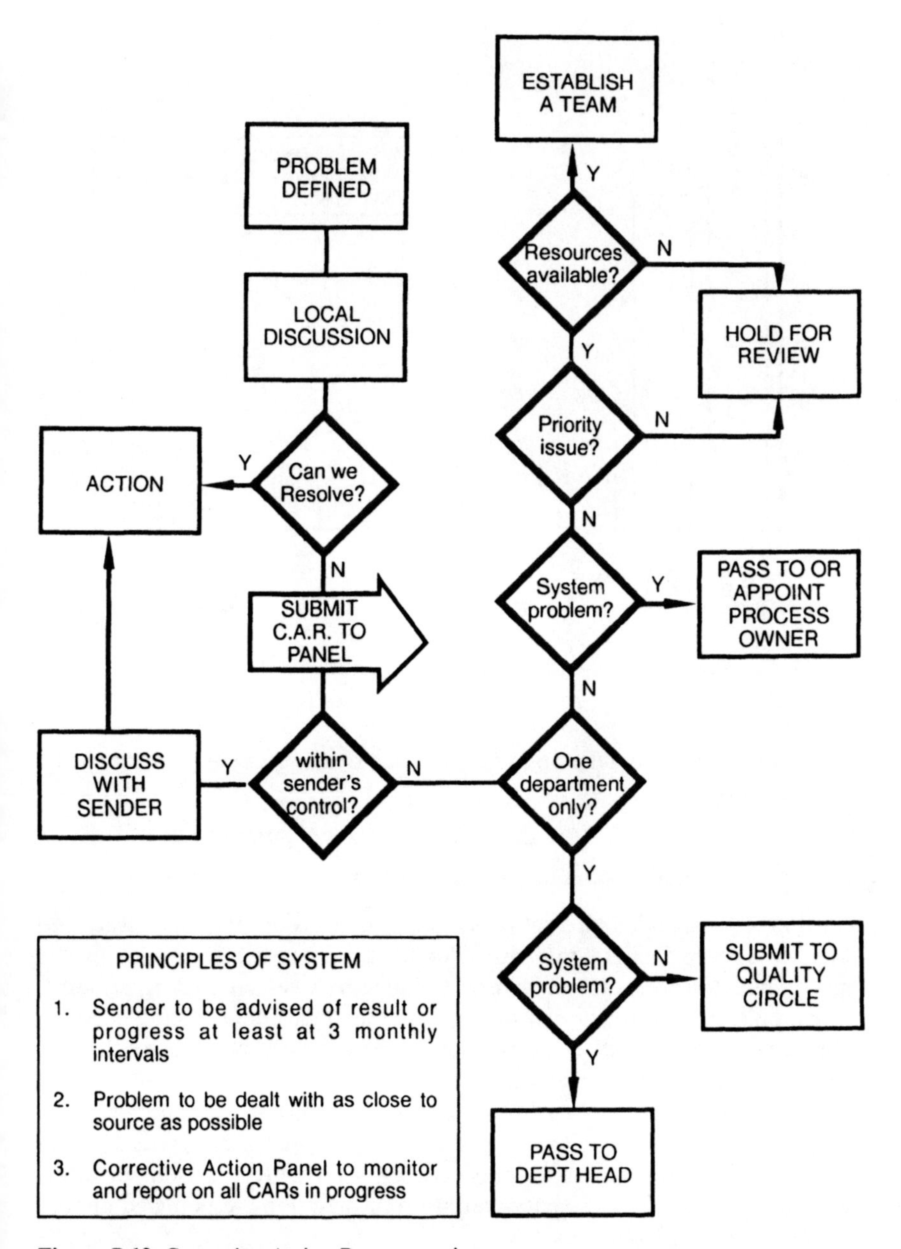

Figure 7.12 Corrective Action Request review process.

It is possible to create an index to assist the use of the Pareto graph, albeit this requires financial data to be available. The Pareto Priority Index

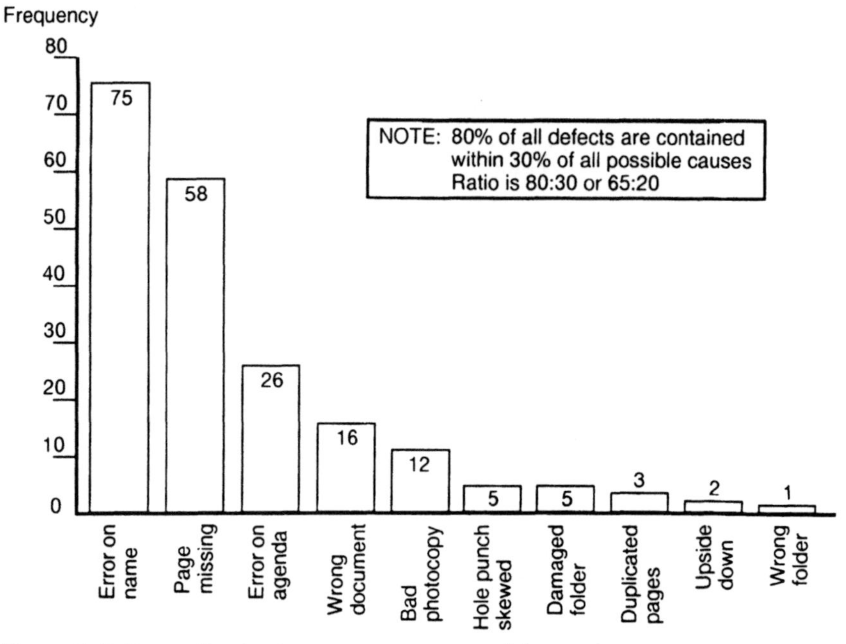

Figure 7.13 Pareto Graph of errors on course materials.

(PPI) is a means of determining best returns from potential projects, using the following formula:

$$\text{PPI} = \frac{\text{Savings} \times \text{Probability of success}}{\text{Cost} \times \text{Time to complete (years)}} \qquad PPI = \frac{\pounds s \times p}{\pounds c \times t}$$

The Pareto method, like brainstorming and decision analysis, described later, is a very flexible tool and can be used at various stages in the improvement process. It can be used to judge between competing projects, rank quality cost and customer or employee feedback data, decide between alternative cause investigations or determine the most beneficial solutions when selecting corrective actions.

The methods described earlier for converting problems into quality costs are valuable in ensuring that a common denominator (cash) is used in evaluating the alternatives. Quality costing also ensures, if used correctly, that the true costs of an activity, variation or omission are being compared and not just the obvious ones.

What is the cost of a missing screw in a packet on sale at a DIY shop? For the shop, probably a fraction of a pence. What is the cost to the customer? Frustration, car journeys, time, etc. What confidence does the customer have in other products if the store cannot even pack screws correctly. A high incidence of failure in areas such as these may lose the customer forever, their faith shaken in the quality of the store. The store managers continue blithely unaware of the problem until they have no customers. If they costed only the material element of the problem, they would never do anything about it. A colleague related to us his annoyance with a European-made car he had bought. Whilst it was mechanically sound, it suffered from an enormous range of glitches – bulbs popping, windows jamming, wiper blades slipping, other electrical instrument failures. Individually, none were catastrophic, nor costly to repair, but collectively they resulted in him swearing never to buy this make again.

CHECKLIST/IDEAS MENU FOR CHAPTER 7

Process control

Redraw your organization as a series of processes and look for opportunities to re-engineer them.

Institute a programme for Departmental Purpose Analysis (DPA) to establish and solidify the understanding of requirements between customer and supplier departments.

Establish a formal process design review procedure and incorporate this into the auditing system.

Assign process ownership to key directors and functional heads and operate a system of certification of key processes for control (consistency) and capability (conformance).

Establish cross-functional Process Improvement Teams between traditional problem interfaces such as finance/personnel or sales/production or marketing/design/procurement.

Teach all managers and employees the difference between common and assignable causes of variation.

Review opportunities for the use of SPC and control charts on key processes.

Encourage the use of checklists among administrative employees.

Problem identification

Set up an error-cause removal system using forms or electronic mail entry. Suitable forms or data entry screens to aid problem definition should be developed by the steering group member responsible for corrective action.

Review your suggestion scheme's place in the quality programme. Does it recognize only a few or most of the employees?

Establish a programme for developing a cost of quality system. Start small – set up a team to collect data and carry out a bellwether project.

Identify, from the critical success factors developed earlier, key measures of performance which can be benchmarked. Establish small study groups on each element to collect comparative data. Do not confine yourself to your own industry but look at the best in the process.

Use customer complaint analysis data to reinforce or challenge your view of key performance criteria. Support this with regular customer perceptions surveys at least at six monthly intervals.

Set up a quality costing system. Start small with a pilot study.

REFERENCES

[1] Juran, J M and Gryna, F M (1988) *Quality Control Handbook*, McGraw-Hill, New York.

[2] Hammer, M and Champy, J (1993) *Re-engineering the Corporation*, Nicholas Brearley Publishing, London.

Making it stick 8

Problem elimination

Permanence

This final chapter describes the steps involved in eliminating the remaining problems after the processes in the organization have been brought under control (in its wider as well as statistical sense). The last of our seven **P**s will then demonstrate how organizations can make the process stick rather than collapse after attention wanes – truly making the transition to a Total Quality culture.

8.1 PROBLEM ELIMINATION

There are several stages in problem elimination, following on from the initial identification and definition of the problem. We defined two stages or journeys in the **P**roblem identification step: the diagnostic and the remedial. Figure 7.7 in that section depicted a path of problem elimination involving these two journeys. So far we have dealt with only the first stage; getting a clear and unambiguous (and non-assumptive) definition. There are four more levels of analysis required to bring the problem from definition through cause identification to solution. These levels are depicted in Figure 8.1.

The Pareto method, discussed earlier as an aid to prioritizing, can also be used in problem elimination by providing a useful diagnostic tool for tracing back the root causes of problems through the process.

In the following example[1], Pareto analysis is used to eliminate one of the most persistent causes of failure in a personal computer configuration:

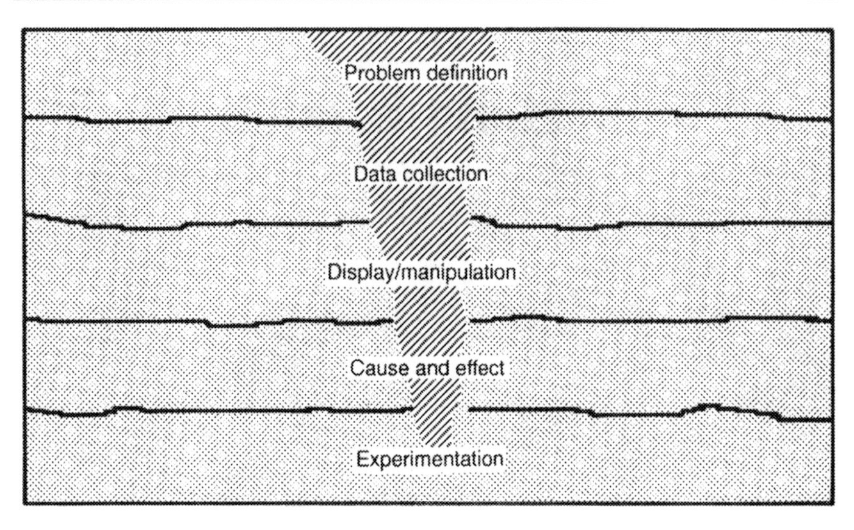

Figure 8.1 Levels of problem analysis.

Product part	Failures
Visual display unit	92
Keyboard	43
Processor	20
Printer	15

The initial location of failures has been identified and listed in order of frequency. The next step is to take the most frequently failed part, in this case the VDU, and analyse the point in the process at which the fault was found. The findings are again listed by Pareto analysis.

Test point	Failures
First part test	80
Configuration test	6
Final inspection	5
Dealer returns	1

As can be seen, most defects have occurred before the VDU is subjected to testing and have been 'built in' in manufacture, or were present already. This should in itself tell the company something about the way it handles quality in production. Clearly there is over-reliance on appraisal with little- or no preventive quality control.

Since most defects were discovered when the part was tested before being put into the configuration, we can then break down the causes of failure discovered at the first part test into the basic units which make up the VDU.

Unit	Failures
Power unit	40
Tube	23
Logic circuit	16
Interface	1

Again we ask – why did the power unit fail? Pareto analysis is applied again to identify the most frequent part involved in the failure of the unit most susceptible to failure.

Part no.	Failures
PX7301	32
PX7302	4
PX7300	2
PX7304	2

By the process of Pareto analysis we have now identified part no. PX7301 as a major cause of failure. We can then apply the principles of problem solving to this part. The cause may lie with the supplier, delivery, storage, processing, or a combination of other factors which cause the part to have failed by the time it reaches the first inspection stage. Tests must be carried out on each of these processes to determine the root cause.

Pareto analysis can be used in all areas of work, not just manufacturing. The marketing department may analyse customers by sales value or complaints handled, accounts by debtor days, typing by reason for returns, or Directors on queries over decisions. For example, in TSB Homeloans, a division of the major high street bank, responsible for processing mortgage applications, a team investigated temporary losses of files and studied the locations where they were most frequently discovered. As a result, they were able to use the Pareto principle to draw up a checklist and process for staff to follow when files could not be located. This led to a reduction in time spent looking for files and a drop from 25% to 2% in the occasions when queries from operations staff could not be answered immediately due to missing papers.

In the personal computer example given, frequency was used to determine the pareto steps but cost or time spent could equally be used if these are considered to be more important measures.

8.2 THE DIAGNOSTIC JOURNEY

The diagnostic journey involves four steps:

1. Deciding the most suitable arena for problem solving; individual, team, department, cross-functional, outside expert
2. Selecting instruments and processes for data collection

3. Selecting appropriate ways to display and manipulate the data for interpretation
4. Working through cause and effect until the root cause or causes are traced

8.2.1 Allocating the problem

A corrective action system is fundamental to the efficient installation of a TQM process. Without this mechanism, there tends to be a scatter-gun approach to the effort.

Benchmarking and Cost of Quality studies, by their very nature, will give rise to a prioritization process. However, the problems which are uncovered through the corrective action or problem reporting system referred to earlier arrive randomly from various sources and one of the jobs of the review panel is to place some priority on these problems and allocate their removal to an appropriate mechanism. This may be reference back to the supplier of the corrective action request if the issue arises from a misunderstanding or lies in the control of the originator. Otherwise the problem may be referred to the appropriate department head, improvement team, quality circle or an individual expert such as an engineer, personnel, finance or R&D specialist.

Those problems which cross functional boundaries will require a team approach in most instances and it will be the task of the review panel to establish cross-functional problem solving groups on high priority problem areas.

8.2.2 Gathering the data

This stage can be split into initial and ongoing steps. Initial data collection simply concerns asking the right questions to allow the problem itself to be properly defined. The **six honest serving men** – who, what, where, how, when and why are used to obtain a statement of the problem which maximizes the facts and minimizes the risks of heading off down the wrong trail.

Some simple data analysis will also usually reveal flaws in any early assumptions about the nature of the problem. The problem statement at the beginning of this section **'Our stationery suppliers are unreliable'**, might be disproved by a control chart. They may in fact be very reliable, but unacceptable. Assumptions that errors are caused by simple carelessness might be thrown into doubt if the data show peaks in certain areas, circumstances or times.

The ongoing collection of data will then be determined by decisions on the likely causes of the problem. Hypotheses as to cause must be proved or disproved; the collection of data is the only way this can be done.

Making decisions as to most likely cause involves the application of appropriate tools and techniques. Brainstorming and the fishbone technique (see Section 8.2.6) are often used by groups at this stage and these are dealt with later.

Instruments for collecting data can be automatic, such as those provided with modern process control equipment or they may be manual systems such as the completion of checksheets or tally boards, or the use of hand held or PC-based electronic storage media. The important issue is not so much that the medium for data collection is available but that there is a clear and consistent process for recording the data and responsibilities are allocated. Individuals must be trained to recognize the data, interpret it and properly complete the record. If the data going into the system are unreliable (this will be determined by the variation in people, methods and materials for data collection), then so will be the results and the decisions based on those results. Quality control of measurement is as relevant as control of any other process.

It is also important to remember that the reliability of measurement is largely the result of people's commitment to carry it out conscientiously. This means that they must be involved in and understand the purpose of the measurement to be instituted. A checklist of questions for introducing a data collection process is given below. **Both of the sets of questions are aimed at prevention of error/failure when introducing measurement.**

Part one – Health check

1.1 Is the organization team or production line based?
1.2 What existing measurement is used? Is this based on quality or quantity?
1.3 Has the management of the group accepted that the staff should be responsible for their own measures and have they demonstrated that to their own staff? If not, how will this be done?
1.4 What elements of the **error-friendly environment** are missing?
1.5 How will these be corrected?

Part two – Planning for introduction

2.1 What is the process to be measured?
(e.g. completion of works order forms)
2.2 What element of the process will be measured?
(e.g. errors on order forms)
2.3 Why was this element selected?
(e.g. it is the simplest overall measure of process performance from the customer's perspective)
2.4 How far down the process has the measurement been set?
(e.g. at the end, i.e. it is still inspection, not prevention)

2.5 How will data be collected?
(e.g. by fault location diagram marked by staff receiving the forms)

2.6 Who will be responsible for collecting the data?
(name or position of person)

2.7 Have they been trained in the purpose and use of measurement for continuous improvement? If not, what action needs to be taken?

2.8 How will they and their team be involved in setting up the system?

2.9 How will the data collected be displayed?
(e.g. by use of Pareto chart)

2.10 What chart definitions (labels) are to be used?
(e.g. x-axis = type of error; y-axis = frequency of error)

2.11 Who will be responsible for plotting the information on the chart/graph?
(refer again to questions 2.7 and 2.8)

2.12 Who needs to know about the results and how will this be communicated to them? (List each customer for the data and the method of communication – e.g. commercial clerks – by team meeting each week.)

2.13 What is to be done with the data?
(e.g. highest frequency error types will be targeted and plotted in order to pinpoint sources of variation and nature of causes)

2.14 Who will be responsible for taking corrective action?
(e.g. commercial manager)

2.15 From the labels you identified in question 2.10, draw a measurement chart for the process element you have chosen and present this on an overhead transparency or flipchart sheet in the plenary session.

8.2.3 Displaying and manipulating the data

Once data have been gathered in rough (input-structured) format, it needs to be interpreted and rearranged in order to allow conclusions to be drawn or hypotheses to be developed (output-structured). The Pareto format is useful for decision making. Other data displays are more useful for specialized purposes; the tool used must be fit for its purpose. It is not the intention of this book to detail every commonly used display method, but the main approaches and their uses are detailed below:

Histograms Illustrating variations in a population or sample at one point in time

Bar charts	Displaying relative quantities or make-up in blocks
Pictograms	Like bar charts but using pictures, such as spanners or cars to represent quantities
Pie charts	Showing divisions of the whole – e.g. between costs of conformance, non-conformance and essential costs
Time series graphs	Displaying trends of results; e.g. frequency of defects against weeks or months
Control charts	An enhancement of a time-series graph illustrating variation over a time period and identifying common and assignable causes, by the use of statistically calculated control limits
Multi-vari diagrams	Illustrating ranges of variation in parts or multiple parts
Defect-grams	Pinpointing areas of defect on a part such as a printed circuit board

8.2.4 Finding the root cause

This involves the accumulation of suggestions about potential causes from the knowledge of those involved in the process under consideration (brainstorming) and a structured approach to isolating the true root cause (fishbone technique).

8.2.5 Brainstorming

The term **brainstorming** comes from a technique developed by Alex Osborne in 1939 to help his advertising agency come up with new and creative ideas for selling his clients' products. The technique is useful at several stages of the problem-solving process; anywhere in fact where a list of ideas are required. These may be ideas about likely causes, potential solutions or potential effects of solutions (the **what might go wrong** question).

Through his experiences in using the method, and those of others after him, it is possible to define a number of key principles for using the technique effectively. These are:

1. The group size should ideally be around four to eight people who have some knowledge of the issue involved.
2. Participants should have some prior knowledge of the objectives of the meeting so that they can think of a few ideas in private first. If

this is not possible, 10 minutes quiet time should be allowed before the brainstorming begins for participants to think.

3. All ideas should be recorded, preferably on a flipchart, no matter how implausible.
4. The first part of the session should have no discussion of the merits of individual contributions but participants are encouraged to suggest improvements to ideas through **idea building** (i.e. **that's a good idea but it would be better still if we did this** . . .)

There are two ways to collect ideas. Osborne's original method involves everyone in the group being asked in turn to contribute one suggestion until all ideas have been recorded (the circle method). The alternative is to allow free contribution (the open channel method). The latter requires more control to ensure there is no criticism or evaluation of the ideas and that the session is not dominated by one or two people and that input is received from the quieter members. Where people are unfamiliar with the technique, it is beneficial, even though some would say a little wooden, to stick to the circle method as this ensures everyone's contribution is elicited and removes some of the fear of making a poor suggestion. A person **passes** if they have no idea when their turn is reached but they continue to be asked in a circle until everyone is passing, since an idea may occur as a result of an idea suggested by another participant.

8.2.6 The fishbone

Once ideas have been recorded they may be ordered into generic headings using the **fishbone** or **cause and effect diagram**. This technique, sometimes called the **Ishikawa diagram** after its late inventor, is used to organize the ideas about potential causes into a logical structure. Wherever, in a process, there is a variation from the planned outcome, the fishbone technique can be used to isolate and identify the potential causes of that variation.

The process of creating a fishbone diagram is as follows:

1. A short form of the problem definition (i.e. the effect being investigated) is placed in the box at the head of the fishbone.
2. The brainstormed list is used as the source of the information with the opportunity being taken at this stage of eliminating duplications and weeding out the ridiculous, impossible or very unlikely or insignificant factors. As these are group ideas rather than the property of one person, there is less (but not none) of a tendency for people to insist on their ideas being carried across.

3. Suitable generic headings for the bones of the fish are chosen by examining the brainstormed list. The old stand-by headings are the four Ms – manpower, methods, materials and machines. Other often-used headings are environment and policies. The latter should in our view be a feature of all fishbone diagrams because the lack of a clear policy is often at the root of organizational problems.
4. Each acceptable cause idea is then listed under one or more of the headings. It is not too important which heading an item goes under; if in doubt, put it under both. What is important is that the cause must be carefully and clearly recorded. Since most problem-solving activity is done over a lengthy period, which may involve several days away from the group and changes of team members, a lot of time can be saved by writing the causes out in detail rather than relying on cryptic words or phrases. This avoids the group having to constantly ask **what did we mean by that?** at each meeting reviewing their fishbone.
5. At this stage a fishbone diagram will contain a lot of information about the problem. An example is shown in Figure 8.2. Sometimes, a certain amount of obvious corrective action may fall out of the process of discussion, but eliminating the root cause will require further data and analysis. Two alternatives exist here. One is to take

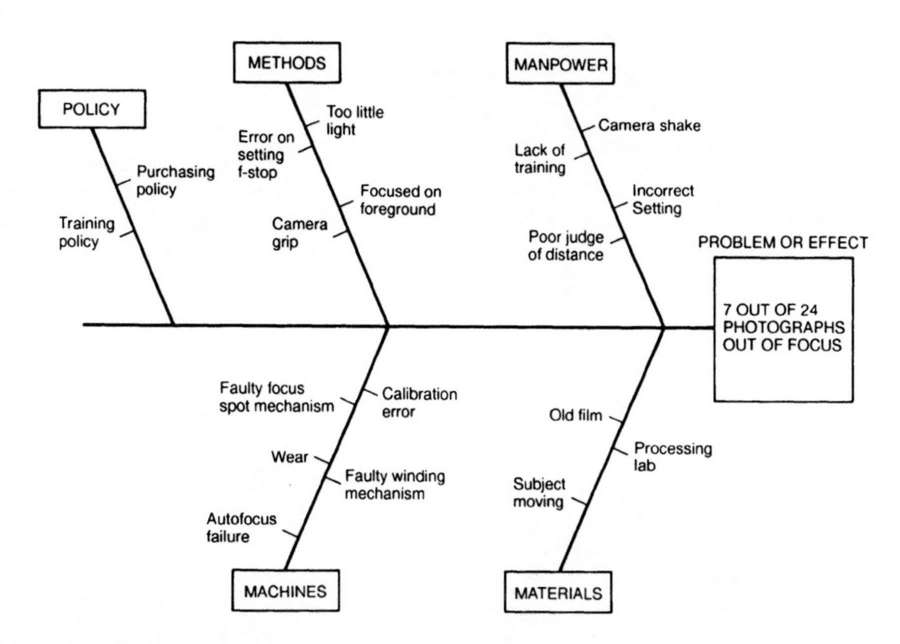

Figure 8.2 Fishbone diagram of photograph focus problem

one of the bones of the fish and break this down further by asking the question why again for each of the influences on that bone (see Figure 8.3). The other is to take a specific cause and isolate this as the effect in the head of a new fishbone. This process of constantly asking why can be followed until the true root cause is reached. Usually two or three backward tracks through the fishbone will start to reveal root causes. One of the phenomena is that the initial **why** questions will result in a range of narrowly focused specific causes, whilst the further questioning of underlying causes will often reveal overall policy or management flaws. For example, the investigation of damage to a fork lift truck might initially identify the cause as driver error. However, the fishbone forces us to ask what causes driver error. We might suggest carelessness. What causes this? Lack of motivation? What causes this? Management behaviour, work organization, reward systems. Blaming the worker is now seen to be unproductive. The root causes lie in our systems. Prevention of future problems by true corrective action on the root cause will eliminate that problem in the future. Fixing symptoms by blaming the person will only result in frustration and resentment.

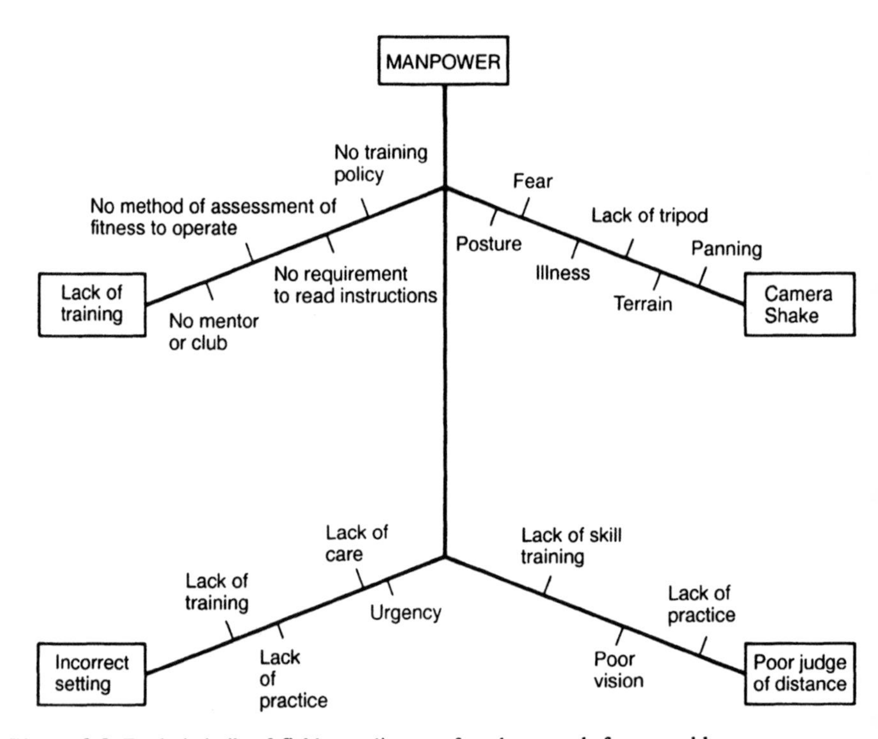

Figure 8.3 Exploded rib of fishbone diagram for photograph focus problem.

An alternative approach to the fishbone is the use of a **why–why** diagram. As the name implies, the tool backtracks the cause and effect process through to its root cause. Unlike the fishbone it is a single track process, concentrating on the most important element at each stage of cause and effect equation.

The example given on pages 236–237 relating to a personal computer configuration and utilizing the Pareto principle could have been drawn as a why-why diagram as shown in Figure 8.4. Brainstorming is used at each level in the diagram to identify the next level of causes and then the most likely or frequent cause is selected using data or a logical decision or voting process. Note how the diagram is used initially as a fact-finding tool to narrow down the problem and then as a logic tool to establish causes.

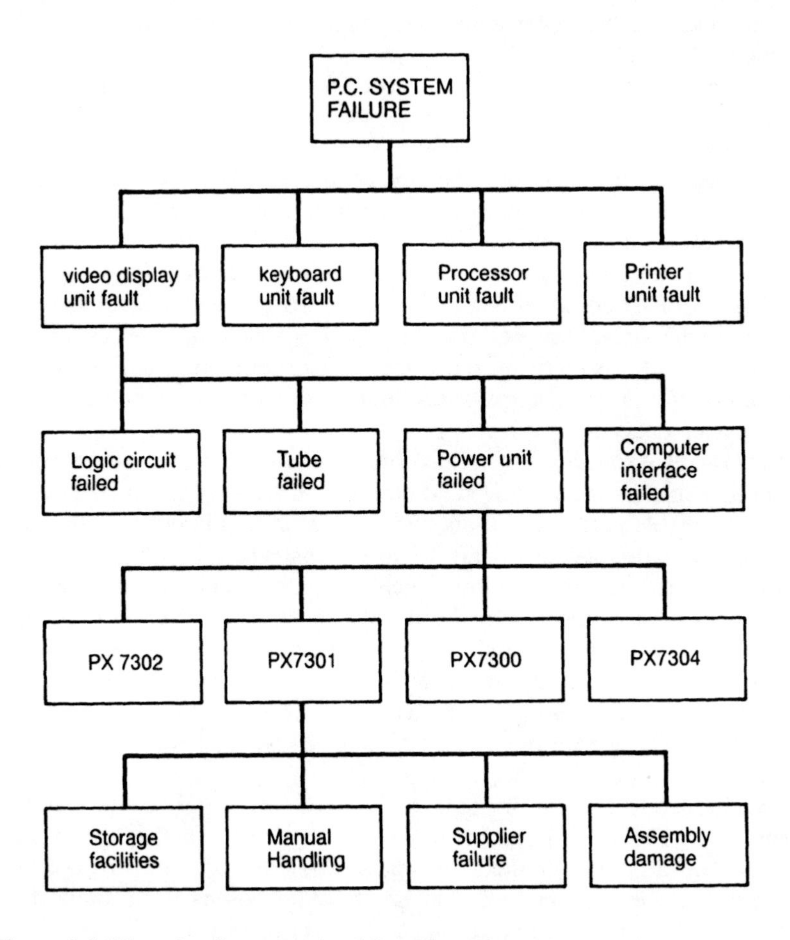

Figure 8.4 Why–why diagram (using PC configuration example).

8.2.7 Identifying the true cause

Don't be led by quantity of suggestions alone in deciding which bone of the fish or which potential cause to explore further. Gather data and use the Pareto principle to identify the most frequent or costly. If this is not possible or is itself too costly or time consuming, a less acceptable alternative is to allow the group to choose the most likely issue through some logical process. Once again the process is important in ensuring we limit the influence of guesswork, generalizations, assumptions and the loud voice.

There are several models around to help groups through decision making. Kepner and Tregoe[2] provide a valuable model of decision making through the use of a ranked and scored matrix where all potential choices are compared against the criteria they need to meet in order to determine the optimum choice.

The criteria the decision has to meet are listed on the left hand side of the matrix. These criteria are then classed either as essential or desirable. Those which are not essential may be weighted for importance. Then the options are listed across the top of the matrix. The next step is to eliminate or adjust those options which do not meet all of the essential requirements for the decision. This leaves a smaller number of ideas which meet the basic requirements for the decision. These options are then in turn scored on a scale of 1 to 10 against each of the criteria, with 10 = fully meets criteria and 0 = does not meet criteria at all. Then multiply weight by score for each criteria and total the scores for each option. The option with the highest overall figure then becomes the tentative decision.

The tentative decision is then rated for its possible adverse consequences on another matrix and reviewed. If some of these adverse consequences are unacceptable in terms of their seriousness or likelihood, another high scoring alternative decision may be tested instead.

Simpler methods are to allow each group member to rank the best causes and then choose from those with the highest overall rankings. Where these are few in number, paired comparisons may be made in which each cause is compared against each other one individually and a score of zero and one given to each pair in turn. At the end of the pairings, the cause with the highest number of ones (score) becomes the chosen cause.

An alternative method using a half-matrix for a paired comparison is shown in Figure 8.5. All alternatives are listed on the left in any order. The same list is then written across the top, leaving out the first item in the list since you do not have to compare an alternative with itself. In the cell intersecting each pair, record a score using the following rules:

- Score 2 if the row is more important than the column. Score 1 if they are equal and 0 if the column is more important than the row.
- Add up the scores in each row and record these to the right of the chart.
- Add up each column but reverse the scores (i.e. 2 becomes 0, 0 becomes 2, 1 remains 1) and place them against the alternative on the row to which they belong (shown in italics).
- Add the two scores together on each row to give a grand total. Note that the first item down the side and the last item on the top will only have one score.

You now have a rough rank order of priorities. The alternative with the highest score may be chosen.

	2	3	4	5	6	7	**Row Total**	**Final Total**
1. Spot mechanism	2	2	0	2	1	2	**9**	**9**
2. Camera Shake	*0*	2	1	2	2	2	**9**	**9**
3. Faulty film		*0*	0	1	1	0	**2**	**2**
4. Moving targets			*5*	2	2	2	**6**	**(11)**
5. Wrong f-stop				*1*	0	0	**0**	**1**
6. Poor judgement					*4*	1	**1**	**5**
7. Light conditions						*5*		**5**

Figure 8.5 Example of a paired comparison on photographic focus fault causes using a half-matrix.

8.2.8 Demonstrating relationships

Sometimes cause and effect are not directly linked but have an indirect relationship. Smoking and heart disease is an example. Smoking will not in itself automatically cause heart disease but it will significantly increase the risk of it occurring. There is an indirect causal link. These links can be traced through the use of a scatter plot which takes two variables and compares the instances of occurrence against one another. The suspected cause is always given the horizontal axis with the suspected effect on the vertical. Data are recorded and points are marked on the graph to enable a **line of best fit** to be drawn to establish any tendency for the two variables to move in tandem. If the variables tend to rise together there is said to be a positive relationship. If one rises as the other falls the relationship is said to be negative. This relationship is known as correlation and can be measured on a scale of -1 to $+1$. Scores of 1, indicated by a straight line, show an exact connection. A score of zero shows no connection. This would be visible on the graph as a series of randomly scattered points with no obvious best fit line. Figures 8.6 to 8.9 below show the four main types of correlation.

The scatter plot is a very useful technique for proving relationships but it is also one of the most dangerous tools around in problem solving. It is

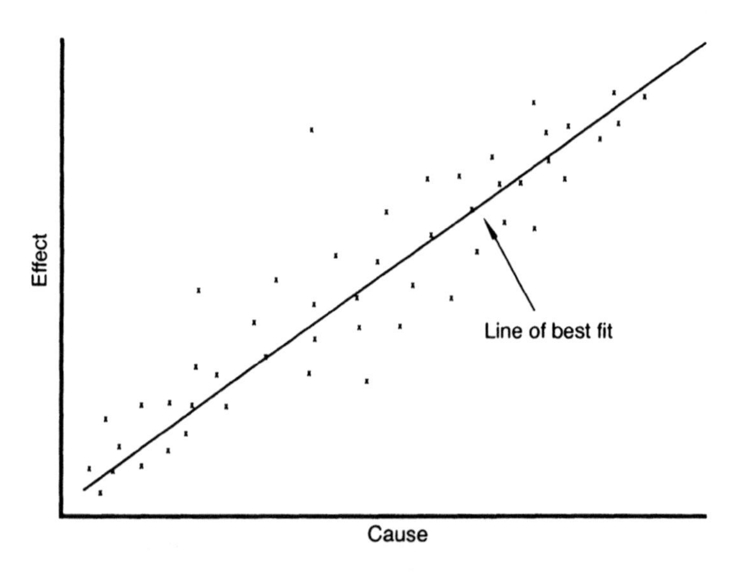

Figure 8.6 Pattern indicating positive correlation.

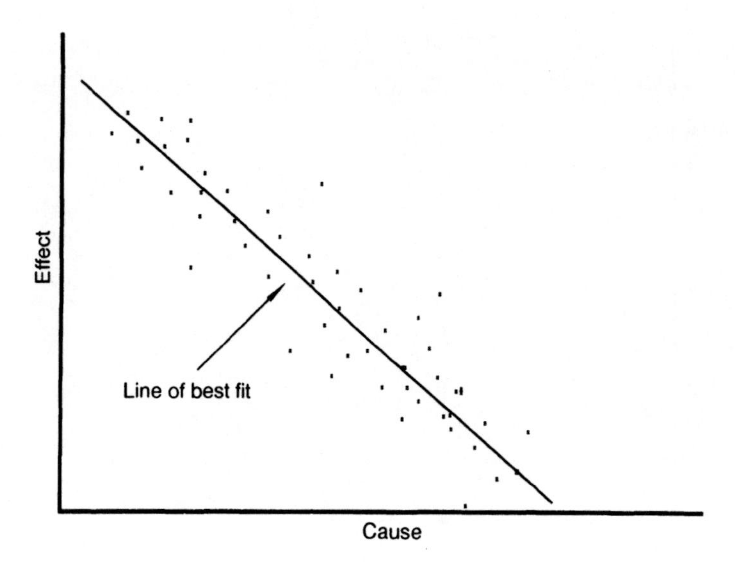

Figure 8.7 Pattern indicating negative correlation.

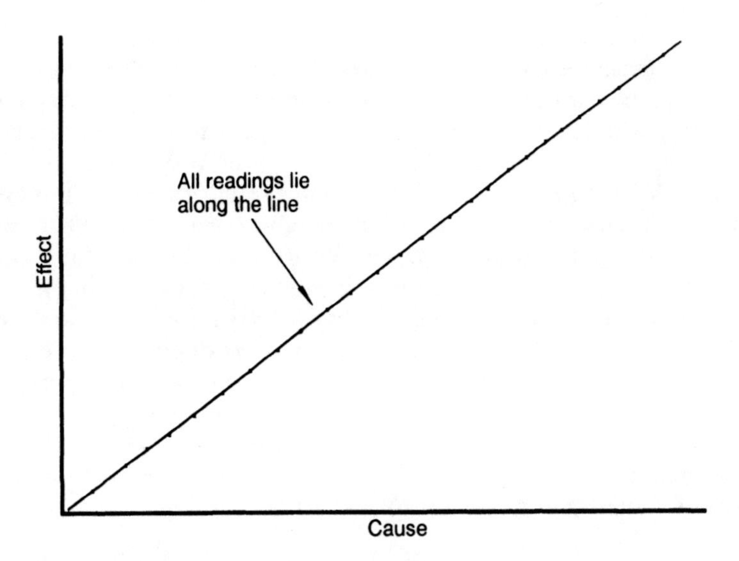

Figure 8.8 Pattern indicating absolute correlation.

very easy to shoot yourself in the foot with it. The main danger is in trying to ensure you are comparing cause with effect and not two effects of an

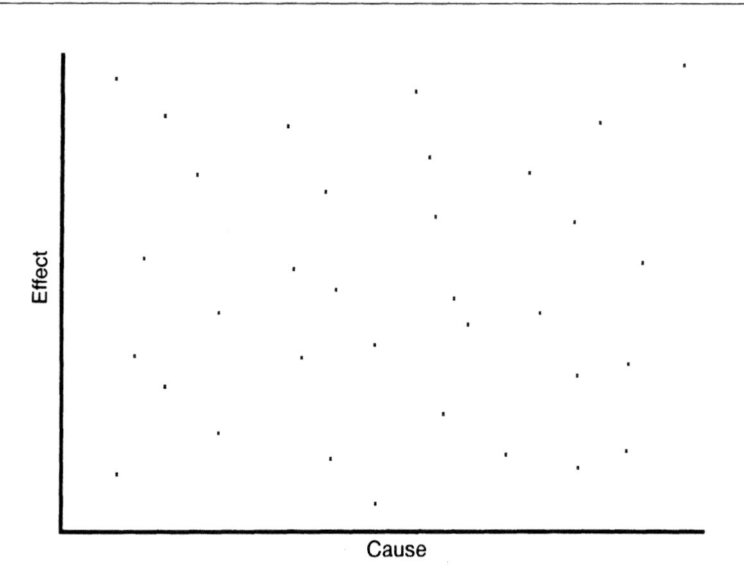

Figure 8.9 Pattern indicating no correlation.

unidentified cause. For many years there was debate about whether there was a third cause, stress, which gave rise to both heart disease and an increase in tendency to smoke. Although the link has now been positively proven it emphasizes the dangers in making assumptions about cause and effect even when we think we have a lot of evidence for it. One of Joe Juran's favourite jokes is about a person who goes to a bar and orders a whiskey and tonic. He has six of these and wakes up with a hangover. The next day he has six gin and tonics with the same result. He repeats this cycle for the next three nights trying rum and tonic, vodka and tonic and brandy and tonic. On the following night he returns to the bar and says to the barman – **skip the tonic this time – it always gives me a hangover**.

Another problem is in assuming you have identified the root cause. For example a study of packing errors and the number of agency staff in use may show a positive correlation. However, the root cause may lie in the amount of training given to people. Whether they are agency or full time may be irrelevant. It may simply be that full time staff are given more training than agency employees. The link therefore is between training and errors rather than type of staff and errors.

8.3 THE REMEDIAL JOURNEY

Three stages are involved here – generating the solution, testing it and implementing it. We deal with each in turn.

7.8.1 Generating solutions

The greatest failing at this stage of the process of problem elimination is in finding a solution which truly removes the root cause and stops it from ever being able to happen again. Prevention of the problem is what we are seeking, not temporary fix or an additional checking or appraisal process.

The difficulty with prevention is that most people don't understand what it is. They think that inspecting and testing is the right way to do things because this **prevents** the error getting any further. In this sense it is prevention, but it doesn't stop the error from occurring in the first place. Take a word processing spell check – it prevents a spelling mistake getting from the electronic source to the printed page but it doesn't prevent it from happening. It would be better to only employ people to use word processors who could spell. This would probably also reduce grammatical and other errors which spell-checkers cannot stop.

Two examples of prevention in a local seaside authority: a water tap by

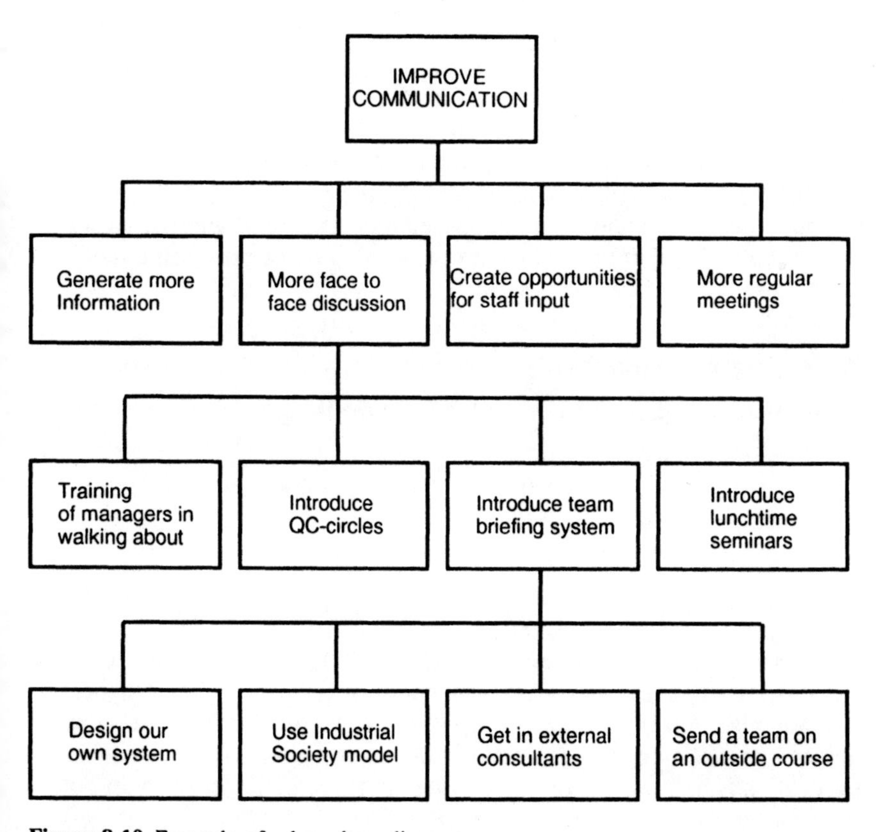

Figure 8.10 Example of a how–how diagram.

a set of beach huts is regularly broken by children swinging down on the tap. It has to be changed about six times a year. The obvious cause of the breakage is misuse, but the root cause lies in the design of the pipe feed which employs four joints. It would cost three times the fixing cost to permanently put it right. The redesign would have paid for itself in one year alone but the council still has a weekly inspection and replaces the tap several times a year – another example of a common cause – a control chart on tap breakages would undoubtedly show a stable pattern of breakages with some taps above and below the control limits. Study of the design of these taps would probably reveal the best design to prevent breakages.

The second seaside example is in beach erosion. Millions of pounds will be spent on new works which could have been avoided by better preventive maintenance, better design and construction of breakers and better materials. These would also have lowered maintenance costs in the long run, as well as saving capital expenditure.

Brainstorming and the decision making approaches discussed under root cause identification can also be used to generate ideas for solutions. Another useful technique is the how–how diagram, which is similar to the why–why diagram shown in Figure 8.4. The difference is that instead of looking for causes the method generates alternative solutions. A solution statement such as **reduce queues at counters** is placed in the end box and alternative ideas are listed below it. The best idea is then chosen or selected through a logical decision-making method. The question of how this action is then achieved is then asked and a range of alternative ways of accomplishing it are listed in the next level. An example of a how-how diagram is shown in Figure 8.10.

8.3.2 Testing solutions

All proposed solutions must be tested for three things:

1. that they work
2. that they truly prevent the problem rather than shift it or kill it by checking it out of the system
3. that they do not cause new trouble elsewhere

The first aspect can be discovered by continuing the measurement of the results until no recurrence is shown.

The second aspect requires a flow chart of the process to be drawn to demonstrate the means by which prevention is achieved. The existence of any appraisal loops indicates an opportunity to further improve performance by redesigning the process to build quality in at an earlier stage. As prevention is built in, the need for inspection at later stages of the process diminishes, saving money for the organization into the bargain.

The third aspect is also concerned with prevention. It is the opposite of the trial-and-error mentality. Any solution will be a change to the status quo. It does not happen in isolation. It will have effects elsewhere. Getting the solution right first time depends on our knowing the effects of the solution. The fishbone or the why–why diagram can be used in reverse here, with the change being placed in the head of the fish or why–why diagram and questions asked through brainstorming on what effects the change will have on different areas, using the bones of the fish or the subsidiary levels of the why–why diagram to define the areas of effect rather than the source of causes.

Failure mode and effect analysis

This technique has long been used in design for manufacture as a means of preventing potential problems in new designs. It can also be used as a diagnostic tool. A new product design is studied and its various failure modes identified. A failure mode is defined as the circumstances under which something might go wrong. A failure mode for a saltpot for example might be hole blocked or refill cap worn. Each of these modes is then studied for what their effects are on the product; e.g. salt will not come out or cap may fall out in use, spilling salt. The causes of failure are then identified, studied for their criticality and preventive or contingency action taken. For example, the saltpot may be redesigned to incorporate a silica lining or a stronger material for the cap.

Criticality is determined by looking at the probability of failure, the seriousness if failure occurs and the likelihood of the failure being spotted before the serious effect occurs.

A similar technique in general management and problem solving provided by Doctors Kepner and Tregoe is known as Potential Problem Analysis. As with FMEA, this involves asking the question **what might go wrong**? All potential problems are identified and rated as to probability and seriousness and an index created. Those with the highest overall ratings, or those where the seriousness rating is very high, are then subjected to an action frame which asks:

- Can the problem be prevented?
- If not, do we need an alternative approach which can be employed if the problem does occur? (a contingency plan).
- How will we find out if the problem has occurred? If it will not be immediately obvious, or too late for the contingency action to be brought in, then a monitoring or feedback loop is considered.

As an example, we may have a training course to run in Manchester. One potential problem is that we are late due to a traffic jam. A preventive action would be to travel the night before and stay over at the venue. If this was not possible, a contingency plan might be to have an alternative tutor close to the venue and a car phone to allow the tutor stuck in traffic to phone in his problem. Monitoring would be by virtue of a travel plan with times when certain points should be reached. Failure to be at Birmingham 2 hours before the course was due to start would trigger a phone call so the contingency plan could be activated.

8.3.3 Implementing the solution

Implementing the solution can be improved by an understanding of the dynamics of change. This is again a preventive approach since it is aimed at ensuring that any change or new system is right the first time. There are several models available to help us. Kurt Lewin's planned change theory sets out three stages in a change process:

Unfreezing existing behaviour

Changing the process/behaviour

Refreezing by rewarding and stressing the changed process or behaviour and its results.

Lewin's approach suggests that actions must be taken to destabilize existing behaviour before action is taken to change it. Once the change has been made, the new system or behaviour needs to be reinforced by management.

Paul Gleicher provides a variation on this theme which has some advantages over Lewin's model. This is described in Chapter 5 under the section on communication and awareness as a formula: $C = (ABD) > X$.

The brackets in Gleicher's formula are symbolically important when it comes to implementing solutions. What they stress is that the factors are mutually dependent. **A zero score in any of the three will render the whole change process impotent.**

In a work setting the introduction of, say, a time recording sheet, will require:

A. A realization of the need by the users – e.g. control of costs, hassle to get materials issued, etc. In effect the organization must sell the need to the staff. The Cost of Quality can often be used as a device to demonstrate the need for TQM.

B. An ability to see how the system will operate – the organization might arrange to show the staff another site where the system works or demonstrate the use of the form by management example or the introduction of a pilot scheme.

D. The knowledge of how to get there – instruction and guidance on the use of the form. User guide notes and careful design of forms might help.

These factors must then outweigh the costs of change – **from the point of view of the people who actually have to fill them in!**

Psychological costs may include poor literacy, suspicion of management's intentions, concern over job security, etc. Practical ones may be dislike of paperwork, finding the time or having a clean area to do the job in. The *A* and *X* factors of Gleicher's formula could be said to represent positive (need for change) and negative (reasons not to change) forces to which another management technique lends itself: that of force field analysis.

8.3.4 Force field analysis

This technique regards any given system, human, operational, technological or psychological, as being in a state of equilibrium with various driving and restraining forces, just as the weather is the result of opposing high and low pressure areas. If high pressure increases, the low pressure area is moved over and better weather arrives, and vice versa. The various forces at play in an organization are complex but can generally be defined as falling into a number of categories. There are forces inherent in yourself (personal), relationships with others, the external environment, technology and methods. Figure 8.11 shows a force field analysis for a smoker who wishes to give up.

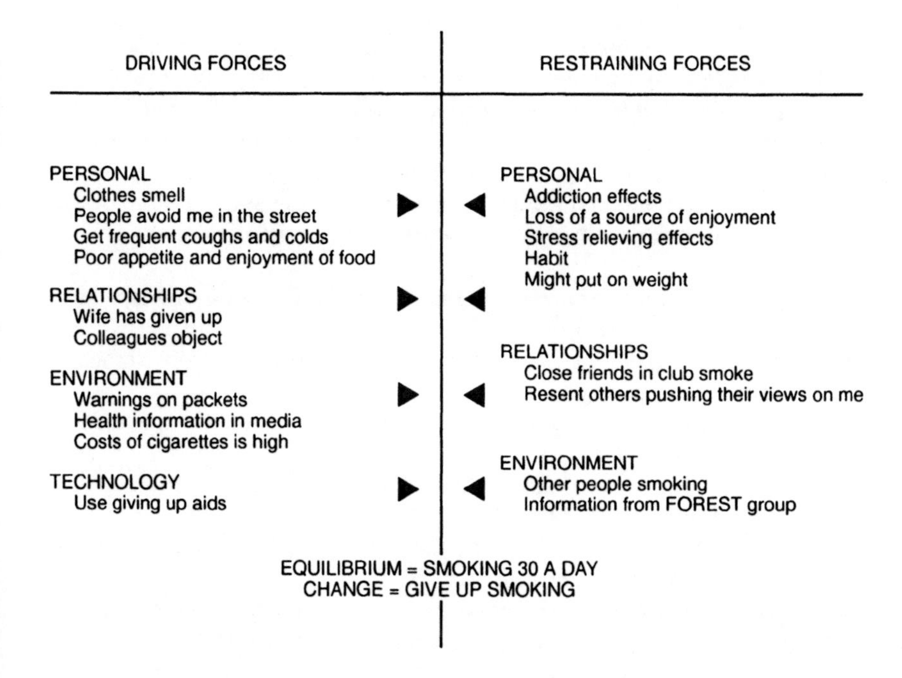

Figure 8.11 Force Field Analysis for smoker wishing to give up

If the smoker wishes to change his current habit, some change in these forces must be caused to occur. For example, he could diet at the same time to reduce the fear of putting on weight, or avoid going to the club. He could make a conscious effort to read up on the health effects of smoking or find an alternative method of stress release. Each of these actions would either decrease the restraining forces or increase those acting in favour of the desired new state.

8.4 PERMANENCE

The seventh and final P for Permanence emphasizes three significant points. Firstly, that quality improvement is never ending. Secondly, that changes have to be made to stick. Thirdly, that control of processes and suppliers will always require attention. We will deal with each of these aspects in turn. Permanence is the step through which all those who seek to be regarded as world class must travel if the culture of TQM is to be realized.

8.4.1 Never-ending improvement

This, above all else, distinguishes TQM from Quality Assurance or any model or award for quality. The attempts to set standards for TQM will never capture the essense of the culture – a deep-rooted and constant desire within everyone in the company to do things better. Management's problem is one of maintaining this philosophy when the initial enthusiasm has worn off. The keys to this are in many ways contained in the earlier pages of this book – as Crosby says in his fourteenth point – **Do it over again**.

The main elements required to maintain progress and get the new way of working embedded in the woodwork of the organization are:

- The continuing use of measurement within the structure of teams based on the natural work group
- The regular addition of new tools and techniques and refreshment of skills with the original ones
- Building TQM practices into the content of the induction programmes for new staff
- Defining practices associated with TQM in management competencies and job descriptions
- Making sure that all employees have personal training/development plans and that these are regularly reviewed
- Managers are trained and competent to apply coaching and counselling skills as well as their more traditional management roles
- Managers' objectives and goals should be reviewed and based on qualitative as well as quantitative measures, including their team

contributions, assistance to teams from other areas and their development of their own staff
- All employees should be able to measure their own performance and encouraged to do so
- Training plans and budgets should be regularly reviewed at senior management level
- Systems are maintained for formally recognizing good team and individual performance. These systems should be supported by managers **walking the job** and understanding the value of recognition by seeking out opportunities to recognize their staff both personally informally and formally with their peers
- The reward system continues to provide support to the business of recognizing all employees' contributions to the improvement process. This may mean changing its basis regularly to avoid **expectation syndrome**
- Regular management reviews of the cultural change process should be held, advised by the results of regular surveys of the staff's perception of the culture
- New goals should be set through at least annual reviews of the Mission and CSFs and then deployed to all staff through techniques such as Business Process Re-engineering, Department Purpose Analysis and Policy Deployment.

The life of the shadow structure

The culture change process is achieved partly through the use of a shadow organization structure as explained in Chapter 5. Many people ask us for how long these structures need to be in place. The answer is the same as that for the person with an injured leg using a crutch – for as long as you need it to help you walk. People take different lengths of time to recover; organizations take different lengths of time to get TQM into the woodwork. Eventually though, the need for structures such as a steering group will disappear and quality improvement will simply have been transferred to the normal activity of the senior management team and every functional and work-based team in the organization. Quality Circles may be replaced with regular work team meetings. Fast-track projects are replaced by the regular and routine use of cross-functional teams generated by the formal corrective action process, Business Process Re-engineering, Policy Deployment and similar techniques. As long as improvement is on every agenda, it does not matter in the long term in what forums the agenda is presented.

8.4.2 Dealing with redundancy

One of the problems which sometimes arises during TQM implementation, or more often than not, from the results of operating more effectively (see text on Business Process Re-engineering in Chapter 7 and Involvement in Chapter 6) is how to handle redundancy. Although TQM is essentially a growth strategy (i.e. winning more business) rather than a reductionist approach to market loss, there will often be times when the requirement for staff diminishes in a particular area of the organization's operation (the middle and supervisory management role is just such an area and is discussed specifically below). If this reduced requirement comes as a result of continuous improvement projects it will not be long before any enthusiasm for being part of a project team or local improvement team disappears. We have, for instance, as consultants, sometimes come in behind Just-In-Time (JIT) initiatives and been greeted with staff who see TQM as another opportunity to **JIT ourselves out of a job**!

You cannot con employees for any significant length of time (for many not at all) – if their involvement is not built upon openness, respect and regard for their welfare, then the process is doomed to expire after a couple of years at most. That is why many companies have taken steps such as those described in Chapter 6 (Section 6.1.2).

Redundancy though is a fact of life. Employees don't expect their employers to be welfare agencies or charities – they recognize that order levels fluctuate, the need for some jobs changes and financial problems happen. What they expect though is that their employers should look at all possible alternatives before using their job security as the solution to the problem. Security is after all one of the most basic levels in either Herzberg's or Maslow's models of motivation and common sense tells us that if people have no feeling of security they will not give their all and they will certainly not participate in activities which may contain their own demise.

Traditionally, redundancy is seen as the quick and easy way out of a problem for UK business. This approach has led to defensiveness on the part of employees, resistance to change from trade unions and a climate of alienation between workforce and management. Redundancy is too often seen as a people problem rather than a process problem – **it is not the people who are redundant, but the processes they carried out**.

The Total Quality Management approach

Many organizations have statements of their values. This should be the first review point when a redundancy situation arises. What do our values say? For example, concern for the individual should mean that personal rather than purely group or union consultation should be the norm.

The employer operating to the principles of TQM should draw some basic guidance from those principles. The key principles are:

- **Prevention** – how can we avoid problems with the process of dealing with the situation?
- **Cost of quality** – what are the real costs of redundancy in terms of lost morale, expertise, customer confidence, increase in failures, absenteeism, etc.?
- **Communication** – what do people need to know and how will we tell them?
- **Involvement** – how can we take people along with us and get them to take a positive approach to solving the problem?
- **Development** – what chances are there to retrain or use people differently?
- **Measurement** – what are the facts about performance if this is to be used as a criteria for selection of people?

These principles should lead the TQM employer to go through a self-questioning process to try to avoid redundancy, such as:

Questions

Is the work finished or will it resume or be replaced at a future date? If the latter, then will short time working, recruitment and overtime freezes get us past a temporary shortfall?

Are there alternative ways of using displaced people?

- In own function?
- With training?
- Within group or associated companies?
- Trial jobs (statutory four weeks anyway)?
- On temporary projects?
- Through suppliers or customers?
- As temporary staff?
- As contractors?

If all these avenues are impracticable or unavailable, what other support should be provided?

- Payment above the statutory rate? Weighted toward vulnerable groups?
- Notice period?
- Early release, pay in lieu?
- Counselling?
- CV preparation?
- Job search?
- Job shop on site?
- Advertising to other local employers?
- Interview skills training?
- References?

Management delayering

One of the most common effects of the teamwork concept and empowerment inherent in a TQM culture is management delayering. This poses a particular problem for many traditional hierarchical organizations moving toward a flatter management structure. As depicted in Figure 8.12, the need to create an empowered workforce has to be matched by an acceptance that the transition will be managed by, and in many ways controlled by, the very people whose jobs will be reduced or eliminated. This is either a problem which will become a barrier, or an opportunity to release resources to support the quality process and add value to the organization's activities. Typically the roles of process engineer, mentor, facilitator, coach and trainer can be rewarding and exciting new ventures for those among the displaced supervisory and management layers capable of being developed into these alternative roles.

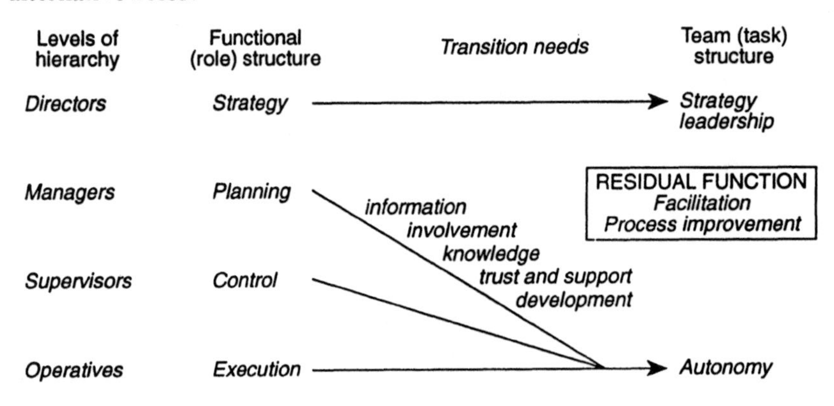

Figure 8.12 Flattening the hierarchy

The transition from hierarchical to flat management needs to be managed and controlled and who better to do this than those who had the responsibility originally. Carefully handled, delayering can be a beneficial experience for all those involved. The key message is to use the opportunities presented by the transition processes themselves to provide breathing space for those who will be displaced so that they can be helped to leave the organization eventually and to provide new roles for those able to adapt to the changed structure.

8.4.3 Making change stick

Permanence's second implication relates to the permanence of the changes made. The **two-year doldrums** is a condition which we have mentioned before. It is the point at which most cascade-type implementation programmes run out of steam. They have finished their company-wide training, the initial projects are completed, often increased workloads have put pressures on them. Symptoms of the two-year doldrums are:

- Steering group meetings are reduced from weekly to monthly or their time shortened or they are merged with the management meeting. The key culture change actions get less and less attention.
- Local managers and supervisors cancel meetings with their staff because of work pressures.
- Project teams' meetings are deferred because one section or another can't release their members.
- New projects are not founded because no effective corrective action system has been set up.

Ultimately, these are all symptoms which are caused by lack of sustained top management commitment. Changes must become part of the system if they are to stick. This means among other things:

- Documenting the new ways of working into the procedures (such as with ISO9000)
- Adjusting job descriptions
- Using the pay and performance management systems to recognize managers' contributions to the new culture
- Conducting internal audits of the effectiveness of the new practices as part of the routine internal quality auditing process
- Using the ISO9000, or other, management reviews to check on the health of the improvement process
- Regularly surveying customers' and employees' views and acting on the results
- Making sure change is planned and managed by using the techniques described in Chapter 7.

8.4.4 Process and supplier control

The third implication of our seventh **P** refers to the fact that while any improvement work is being undertaken the basic elements of quality still have to be managed – permanently. Supplier development programmes and process control techniques have to be **constantly** applied, not just as part of a project or as a task in the implementation of TQM.

Process control has been dealt with in the previous chapter. The methods described in that chapter, once applied properly, need to be extended backwards into the supply chain.

The control of suppliers has been deliberately left until last. It is our belief that organizations should put their own houses in order before they start throwing stones at suppliers. The same basic principle which applies to quality circles is also true at a macro-organizational level, i.e. that if everyone works at improving their **own** processes first there will be fewer cross-functional problems to resolve.

8.5 SUPPLIER QUALITY ASSURANCE AND IMPROVEMENT

The movement in concern for quality is reflected in the attitude of buyers, traditionally viewed as the people whose job it is to turn the screw on price. A survey in 1979 showed only 3 to 4 buyers in 10 held quality equal to or higher than price. By 1988, this had doubled to 8 in 10. This is not to say that cost is unimportant but most buyers today see quality as the first gate a supplier must pass through. Then and only then is price considered as one of a number of parameters.

Supplier control issues fall into a number of categories:

- Selection
- Auditing
- Partnership arrangements
- Training
- Quality improvement
- Involvement in design, often expressed as **concurrent** or **simultaneous engineering**

The basic belief underpinning the supplier control issues in TQM is the idea that you are **purchasing** the supplier rather than their product! We will deal with each area in turn.

8.5.1 Supplier selection

Here are a few ways not to select your suppliers:

- *Yellow Pages*
- Impressive advertising
- They did a good job for so and so
- We know them
- They are part of our corporate group of companies
- They called
- They are cheapest, nearest, biggest, smallest or any other non-quality criteria

That is not to say that the above reasons are not valid – but they should only be used to get a supplier on to a 'possibles' list. (In formal tendering, the requirements may be much tighter for getting on to the list.)

The selection of most suppliers, other than of routine non-critical commodity supplies, should be based on a rigorous assessment of their capability. This capability study should cover several key areas such as:

- **Specification**. Can the supplier meet all aspects of your requirements and are they willing to adjust (tailor) their specifications to your needs?
- **Cost**. Are they willing to show how their prices are built up (including their margin) and what actions are being taken to control or reduce cost?
- **Delivery ability**. How quickly can they guarantee delivery from order? What is their record in meeting these targets? What other sources are available to them if there is a failure?
- **Experience**. How long have they been performing in this field? Is there any evidence of innovation? What access to other customers is there?
- **Technical or problem back-up**. How much technical expertise do they have? Is it local? How quickly will they respond if there is a problem?
- **Resources**. How near capacity are they? How do they cope with peaks and troughs? Is the manufacturing system flexible?
- **Commitment to quality**. Do they have a quality policy? Are they independently assured (e.g. ISO9000)? Do they have a robust quality improvement programme under a TQM or other banner?
- **Evidence of performance and variation in key characteristics of product or service**. Can they provide statistical evidence? Are they willing to let your own engineers or specialists in to view?

General Motors 'Targets for Excellence' programme, for example, assesses suppliers according to their system's reliability and capability in the following areas:

- Targets for excellence
- Continuous improvement
- Management
- Quality - Price - Delivery - Technology

The extent of a GM asssessment of a supplier's capability goes well beyond the normal enquiry range of a QA audit under ISO9000. A few selections from a GM assessor's check sheet reveal the depth of the difference in their approach:

- **Organization** – is the number of levels of management appropriate for the firm's needs?
- **Finance** – review supplier's projected sales figures.
- **Communication** – review of effectiveness of operational meetings and media used for employee communication.
- **Planning** – does supplier have a long-term business plan? Supplier should demonstrate awareness of competition and their analysis of the marketplace. Plan's objectives should be reviewed at least once a year with employees.
- **Training** – should be a budgeted item and have a designated individual to administer it.
- **Employee development** – supplier should have organized programmes

to promote employee participation such as problem-solving teams and suggestion schemes.
- **Employee satisfaction** – review absence and turnover rates. Evaluate employee satisfaction with their work and environment. Review labour dispute history. Is benefit programme appropriate?
- **Housekeeping** – evaluate neatness and cleanliness of plant, facilities and grounds.
- **Health and safety** – are restricted areas of plant clearly designated?
- **Supplier selection** – does supplier use competitive quoting and comprehensive assessment of supplier business practices? Does supplier work with its own suppliers to improve their performance on quality, delivery and cost reduction?
- **Customer responsiveness** – assess systems to address all customer-related issues.

To single source or not to single source?

There are differences even amongst the experts here. Deming sees single sourcing as fundamental to the reduction of variation and improvement of reliability and performance. Single sourcing in the sense used by Deming means the source of the product – i.e. the machine, the crew, etc. – not the name of the supplier, who may have many different internal sources of supply (different sites, machines, crews, thus increasing variability). Juran on the other hand points out the dangers of allowing complacency to creep into suppliers and the growth of the monopoly mentality. One constant, however, whichever view you take, is that in general organizations use too many suppliers and do not manage the relationship to either party's long-term benefit.

This results in a phenomenon of most serious quality improvement processes which is the reduction in numbers, often on a huge scale, of suppliers and the growth of longer-term partnering (discussed later) arrangements. For example, **3M** went from 3500 to 350, **General Electric** from 1600 to 200, and **Dupont Carriers** from 1400 to just 12. **Xerox** cut their production suppliers from 5000 in 1982 to 400 in 1989, **Mitel** have reduced from 500 to 400 with a target of 250 to 300. Among car manufacturers, **Ford Europe** went from 2500 to 900, **Peugeot** from 2000 to 950, **Jaguar** from 2000 to 700, **Renault** from 1400 to 900 and **Austin Rover** from 1200 to 800.[3]

8.5.2 Supplier audits

Having chosen ('purchased') your supplier by careful assessment, the continuing relationship should be one of mutual trust. For this reason, some experts suggest that the formal audit is counter-productive. Deming, for example, suggests that suppliers be required to supply statistical proof of conformance. It can also be argued that a supplier who is already third-party assured does not need to be audited by each and every customer. Ultimately, the question is one of risk-management. If the supply is critical, subject to significant variation outside specification, has a history of problems, or internal or third-party controls are inadequate, auditing may be necessary.

Supplier auditing is distinguished from supplier assessment here as the process of checking on a supplier after selection on to an approved or 'in use' list. The range of enquiry in auditing is narrower than in assessment and should focus on the key performance criteria agreed between the parties. The audit should be a combination of inspection of relevant process performance (e.g. materials storage, defective parts control) within the supplier's site and a review of historical data regarding product or service conformance such as product out of spec or service failures (e.g. late delivery or responses to enquiries).

Auditing can be thought of then as the maintenance process applied once you have chosen to approve (i.e. purchased) a supplier.

8.5.3 Supplier partnering

At its most basic, partnering is simply an expression of the relationship between the parties and represents a move away from the kind of one-sided relationships we have seen in the past where supplier clout has been used by companies like Coca-Cola to restrict competition and where buyer clout has been used by some companies, notably in some parts of the public sector and retail business, to peg price, occasionally at the risk of suppliers' survival.

In its more developed form, partnering represents a much longer-term merger of skills and resources between two or more, usually vertically linked, businesses in a supply chain and can incorporate some of the features discussed below under training and design.

Although it is impossible to absolutely specify what distinguishes a long-term approved supplier from a partnering arrangement, most of the following criteria should be present in the latter:

- An open book arrangement where each party understands the other's product pricing make-up and has agreed margins.
- Pre-contract discussions on product design and manufacturing or service supply processes. Pre-contract here means pre-contract with a third party who is the customer of the product or service provided by the company and its supplier(s) working together (i.e. the partnership).
- Shared resources and costs in areas such as design, training, information technology, publicity and marketing (such as piggy-back advertising, e.g. Hotpoint/Persil).

- Some mutually owned stock. Each has a hand in the other's pocket. Rover and Honda formed one of the most well known of these arrangements until recently.
- Collaborative research into new products and joint development.
- Joint planning teams for manufacturing or service delivery processes.
- High levels of information exchange – both personally between designers, engineers, sales and procurement staff and technically through, typically, electronic data interchange (EDI) on stock levels, ordering, etc.
- Agreed cost and price reduction targets based on joint improvement initiatives – often aimed at product or service interface issues such as stock coding, warehousing, billing and payment, delivery, stocking, etc.
- Development of special expertise by the supplier/partner relating to the customer's product or market.
- Occasional staff exchange or staff working and based at each other's premises.

8.5.4 Supplier training

Most companies only get involved with their suppliers when something goes wrong or when controlling them by means of audit or assessment. However, a TQM company should be looking for scope to apply more prevention-oriented initiatives such as training and familiarization of suppliers with the customer's processes, needs and potential problems. British Telecom, for example, operate a 'Top Suppliers Forum' with their main vendors which involves a regular series of bilateral meetings. Smaller suppliers are invited to presentations at which BT discusses its products, policies, programmes and procedures. They are now introducing programmes to share their training expertise and resources on TQM with their suppliers.

Some of the best known examples of training for suppliers come from the automotive industry. Most of the quality motor manufacturers provide training in both directions in the supply chain: to suppliers on manufacturing and engineering principles, increasingly using SPC; to their dealer networks on maintenance problems, product features and other aspects where there is a direct benefit for customer brand loyalty.

Getting the arrangements right first time depends on effort expended up front through assessment and then through early training. The **induction** of a new supplier is critical – familiarization training for the supplier with the principles and expectations of the relationship through seminars should be provided within weeks of an agreement to purchase.

Ongoing training can then be focused on areas of mutual interest. In one of our middle management workshops with a client planning the introduction of TQM into the company, one of their own suppliers presented part of the course, using the opportunity to get across the message to a wide audience regarding the criticality of hygiene when supplying cardboard packaging to the pharmaceutical market.

'Mutual benefit?' you might say – surely the customer was seizing an opportunity to complain to a wider audience. Not so. The customer certainly stood to benefit from his investment of time but so did the supplier. The success of future work contracts depended on a factor which had not been well communicated to operational staff in a traditionally dirty industry. The attendees on the programme had never really considered the requirements for hygiene until the problem was graphically presented to them through rejects and their own customer's returns brought in by the purchasing director of the pharmaceutical company. The message stuck – this was not just a memo or procedure: this was real!

8.5.5 Supplier quality improvement

Handling of supplier problems also takes on a different flavour. In the past the tendency had been to switch the supplier (an example of failure to understand the difference between common and assignable causes of failure and act accordingly). This approach also resulted in suppliers having no interest in long-term improvement since they could rely no further than the next order or problem. Often they would not even be told the reason for the switch. Single sourcing or dual sourced partnering arrangements, however, do encourage the supplier to work with the customer to improve quality; they have a vested interest in doing so – the reward of long-term order guarantees.

Many companies have established Supplier Quality Improvement Teams (SQITs – a lovely acronym!) to work jointly on issues affecting supply problems. Leyland Daf's assembly plant at Leyland in Lancashire report actions taken with a German supplier with frequent non-conformances. They set up a quality improvement team involving employees of both companies which led to a 90% reduction in non-conforming supplies saving £9000 per month. As a result the supplier was able to hold his prices for two years on the back of the saving.

Awards and recognition of suppliers' contributions and centrality to the quality improvement process also feature prominently in supplier quality programmes. IBM's 'Quality Performance Award' requires suppliers to provide at least one million dollars worth of goods over a twelve month period with 100% batch acceptance and no line rejects or parts scrapped.

8.5.6 Supplier involvement in design

In its advanced form, design involvement is known as **simultaneous** or **concurrent engineering**. It is a formal long-term contractual arrangement, usually between an OEM and one or more suppliers of materials and services, which promises a lengthy period, usually in the region of three to five years, of guaranteed purchases by the OEM in return for an investment by the supplier of their time and resources in the simultaneous design of parts.

The objectives of concurrent engineering can be summarized as:

- Early prototype production
- Reduced time to market of new products
- Manufacturing and engineering feasibility issues reduced
- Earlier introduction of statistical controls
- Fewer design modifications after product launch
- Development of learning resources by both parties for future use
- Sharing of design costs
- Reduction in launch costs
- Reduction in materials costs through value analysis techniques
- Reduction in complaints and problems from the end product user.

The details and methodologies behind simultaneous engineering are too many for this book to cover but it is worth making special mention of one technique generally employed in this process which has revolutionized the design process itself (an example of business process re-engineering in fact) – Quality Function Deployment, or QFD to use yet another TLA (three letter acronym!)

8.5.7 Quality Function Deployment (QFD)

This technique is in essence a design tool. Organizations which have developed a simultaneous engineering process frequently use it as the main design technique. It was developed in Japan's shipbuilding industry but has found wide application, most notably in vehicle and electronics manufacturing. Toyota used the technique to steal a march over its rivals, employing QFD to reduce the time it takes to bring a new vehicle from conception into full production. They reduced their lead time capability to the point where they were a full two years ahead of their Western rivals. In the process they also dramatically reduced the number of design modifications required, shifting the emphasis on to prevention as shown in Figure 8.3[4].

The application of this somewhat cryptically titled process involves the use of multiple level matrices which optimize product design at the initial design stage by integrating the voice of the customer with the voices of engineering, manufacturing, procurement and quality to ensure the design is right first time. This contrasts with the more usual 'over-the-wall' approach where marketing drive the design team who then lob their design into engineering who chip at it to meet their principles. They in turn lob it into manufacturing who do likewise. The end result of this process is internal friction, Chinese riddles, design flaws introduced post-design stage, numerous changes and references back, with the customer usually losing out.

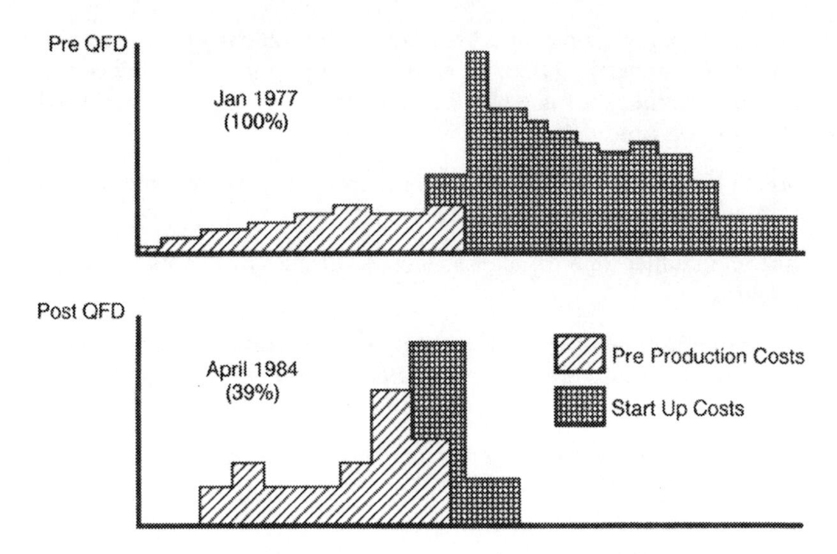

Figure 8.13 Results of Toyota's investment in Quality Function Deployment.

The whole area of quality in design and its associated tools and techniques such as QFD and Taguchi methods is best left to another book as a few examples would not do the subject justice. Suffice to say that many process design reviews may be stimulated by QFD, in an effort to ensure the voice of the customer stays with the product from concept through to production, rather than being optimized out well before it gets there. Its relevance to the **P**ermanence step is that it drives continuous improvement from the stage which ultimately is the one where prevention and getting it right first time really matter - design.

All the techniques described in Chapter 8 depict a company reaching maturity in its journey toward total quality. In particular, the control of suppliers is seen as a later stage because there is one thing you must always bear in mind when setting in train the processes described - **be sure *you* can do the things you are asking your suppliers to do when *your* customers want it from you!**

CHECKLIST/IDEAS MENU FOR CHAPTER 8

Problem elimination

Source training for all senior managers, and the steering group in particular, in a wide range of problem-solving techniques and evaluate the best approaches. Make sure tools are available covering all the steps in the problem-solving process, from definition of the problem to implementing the solution.

Train facilitators to appreciate a broad range of problem-solving techniques and use the knowledge gained to help the steering group define common tools and techniques for use by all employees. Do not overlook the value of SPC in non-production processes.

Establish a detailed training programme in appropriate problem-solving tools for all employees.

Define and budget time for staff to meet to discuss problems within their own teams.

Set up a programme of measurement systems reviews, for all managers, of data produced in their own areas. Focus to be on helping staff make better use of data rather than criticizing results.

Establish a recognition scheme for error-finding, prevention and problem-solving.

Permanence

Use Quality Policy Deployment to make sure management never takes its eye off the ball.

Rearrange the design process to provide for team-based design with maximum customer participation, utilizing tools such as Quality Function Deployment.

Carry out an analysis of suppliers by product or service category and review opportunities for reduction of supplier base, partnering and single sourcing.

Review supplier selection procedure and ensure certification process is clear and based on quality.

Institute a programme of supplier assessment and joint reviews.

Establish a supplier quality training programme.

Set up an award system for suppliers.

Set up joint Supplier Quality Improvement Teams (SQITs).

Hold an open day for suppliers to affirm your commitment to and their involvement in the quality improvement process.

Institute communication systems for suppliers to report quality problems you give them.

Check out all buyers' job remits for emphasis on quality and ensure there is a feedback system to the buyer for problems in use of purchased goods and services.

Review the number of layers of management and supervision – consider the potential for team or cell-based structures which eliminate management layers.

Review the redundancy procedure and policy for opportunities to create **employment** (rather than job) security for staff.

REFERENCES

[1] We are indebted to Crosby Associates for the idea for this example – the case is similar to one used in their educational material for managers known as the Quality Education System.

[2] Kepner, C H and Tregoe, B (1981) *The New Rational Manager*, Princeton Research Press, Princeton N.J.

[3] *London Evening Standard*, 23 May 1990.

[4] Sullivan, Lawrence P, 'Quality Function Deployment', *Quality Progress*, June 1986.

IN CONCLUSION

As we have produced this first revision to *TQM in Action*, the management fashion industry continues to dictate the pace of change in TQM – most of it regrettably in semantics. TQM becomes WCM or VTC or worse, techniques such as Business Process Re-engineering are sold as replacement panaceas for TQM rather than an example of continuous improvement in TQM techniques. More babies disappear with the bathwater.

Since we wrote the original book, the number of organizations claiming to have some form of continuous improvement process in place has grown dramatically. The British Quality Award has been launched and the European Quality Award has developed a new industry for itself. Assessment has become the biggest growth sector in the economy, whether for ISO9000, BS7750, TQM, Health and Safety, Investors in People, NVQs, or the Accreditation of Prior Learning (APL), and for every assessment process there seem to be an army of assessors to assess them. We seem to be flattening our management structures while building towers of hierarchical assessment processes. No one questions the value of assessment itself but there is a danger of it becoming a substitute (and not a very good one at that) for good management these days.

This book has sought to set out a range of actions you can take to establish and maintain a TQM process in your company. Undoubtedly, there will have been things we haven't included which you thought we should have. There will also have been things mentioned which merit further enquiry. You will never find all the answers in one text book. Joseph Juran's QC Handbook is probably as near to that as anyone has come but we defy anyone to class it as a good read. We hope that we have been able to combine a good read with a range of ideas and an approach to managing your programme which you will find of use. Remember though that the search for improvement is never-ending. Treat that statement as applying to everything you do, including your own development as a manager, employee or student and you will always be up there among the best.

A final thought

The trouble with book publishing is that industrial and economic change is much faster than the publishing process – a case for re-engineering perhaps?

We guess it just means we will have to keep on improving too.

PART THREE

Implementing Total Quality Management: A Case Study

Introduction to the case study

This case study describes what happened in GSi Travel and Transportation (UK) between September 1990 and January 1995 in this company's efforts to implement a Total Quality Management Process. It also describes the recollections of some of the members of the Steering Group/Management Team and of some of the staff, looking back with the benefits of hindsight, over the period. These views were expressed during a series of interviews conducted in January 1995.

We believe the case provides a number of valuable insights into the complexities of the implementation process and the amount of time the process requires to get started, to keep it going and to get the most from it. It reinforces the view that management commitment is essential and that a considerable investment of time and money is involved in training managers and employees at all levels throughout the company in order to obtain the benefits of continuous improvement and cultural change.

Managers in GSi Travel and Transportation (UK) would be the first to admit that, despite the efforts made so far, they still have a long way to go and are still learning. We hope this case will shorten the learning process for others and lead them to even better results as a consequence.

ACKNOWLEDGEMENTS

We would like to thank all those who contributed to the writing of this case. In particular we would like to express our appreciation to Greg Coady, the Director of GSi Travel and Transportation (UK) for his time, for allowing us access to company documentation and for permission to publish our findings. We would also like to thank Alistair Grant, the TQM Manager, for the generous amount of time he spent with us, and the other members of the Management Team and staff who participated so willingly in the interviews.

Finally, the case study would never have appeared in print without the help of Claire Stone, Research Assistant, Mrs Lesley Parry, Administrator, both from the Total Quality and Innovation Management Centre at Danbury, and Mrs Veronica Williams, who gave up her weekend to provide secretarial support to ensure we completed everything on time.

Introduction to the company

1.1 INTRODUCTION

GSi Travel and Transportation (UK) Limited is part of GSi Travel and Transportation, which in turn is part of the GSi Group. The GSi Group is one of Europe's largest software service companies with a turnover of approximately £250 million. It operates throughout Europe and North America. The group consists of eight business activities:

- Travel and Transportation
- Motor Trade
- Business Management
- Payroll and Human Resources
- Engineering Facilities Management
- Advanced Technologies
- Marketing and Economics
- Banking

GSi Travel and Transportation is located in Camberley (UK), Paris, Darmstadt, Madrid, Zurich, Amsterdam and Brussels. It is 34% owned by French and German PTTs.

GSi Travel and Transportation are specialists in:

- Reservations
- Clearing systems (BSP and CASS)
- Air and surface freight systems
- Travel agency systems
- Tour operator systems
- Relationship marketing applications

GSi Travel and Transportation (UK) Limited currently employs 270 staff, has a turnover of approximately £8 million per annum and handles 20 million documents annually.

Its mission statement is:

To be the recognized experts in the provision of clearing and information management services for the travel industry.

Its key features are:

- Two major activities:
  - clearing house for settlement of accounts between airlines and agents for the sale of tickets and freight
  - direct marketing sales and customer support
- Ten years of processing UK bank settlement plan
- Located close to road and rail links with major customers.

The company are specialists in:

- Systems and programming
- Communications and networking
- Data entry
- High volume document clearing
- 'Frequent user' club management
- Quality fulfilment.

Some examples of the above are:

Clearing

- Managing the BSP (bank settlement plan) on behalf of the carriers since 1984
- Running the cargo equivalent (CASS) in the UK
- Providing a facility (CARDCLEAR) to allow settlement of internationally issued credit card sales in the currency and bank account of the clients' choice.

This is shown diagrammatically in Figure CS1.1.

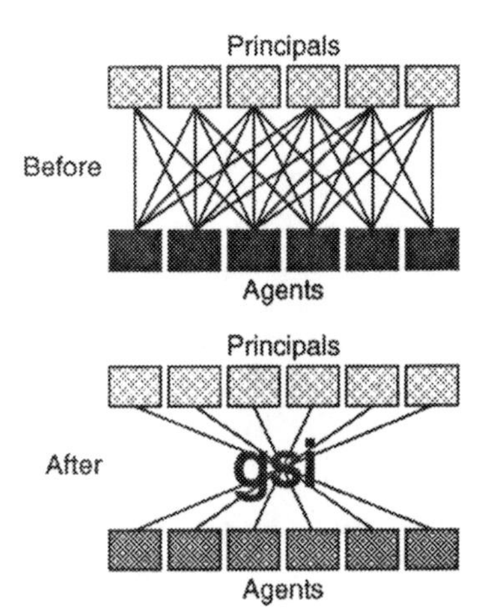

Figure CS1.1

Two further examples are:

'Club' management and fulfilment by GSi

- Operated since 1982
- Customers include:
  - British Airways
  - Qantas
  - Diners Club
- High quality, personal service to club members recognizing their value to the organization
- Specialists in club management – *not* lick, stick and stuff.

Administration and fulfilment

Customer-related services include:

- general administration
- call management
- letter and merchandise fulfilment
- database maintenance.

The organization of the company is shown in Figure CS1.2.

ORGANIZATION

MANAGEMENT TEAM

- ☐ 8 senior managers
- ☐ sets local policy
- ☐ performs business planning
- ☐ promotes and coordinates TQM programme

OPERATING DEPARTMENTS

- ☐ document handling
- ☐ data capture
- ☐ computer processing
- ☐ direct marketing

SUPPORT FUNCTIONS

- ☐ finance
- ☐ training
- ☐ systems support and development
- ☐ marketing presentations and major account management

Figure CS1.2

1.2 SOME OBSERVATIONS OF CHANGE IN THE MARKETPLACE SINCE THE START OF THE TQM IMPLEMENTATION PROCESS

The Managing Director

The marketplace is getting more cost conscious. Potentially more competitors are moving into our geographic area – that is, in Continental Europe. The demand for information faster has increased. There is a hunger to get information more quickly which is reflected in the demands from our customers on us. We have a monopoly in the UK – 100% of the market. Our contract has been extended – it expired in mid-1996 but has been extended to the end of 1998 (i.e. the BSP contract).

These changes reinforce the need for the TQ programme. We have constant concerns that our customers' expectations of us have not been high enough. We have been allowed to get away with unexceptional performance. Our customers didn't like to complain because we are doing a similar job to them. Because they couldn't get it right, we became complacent. Because there were no complaints it was assumed we were okay. It's new ground to ask customers what causes their problems in order to see improvements. Our objective is to want customers to stay with us because we are doing the best possible job.

The TQM Training Manager:

There has been a significant shift in the marketplace from paper-driven to electronic-driven systems. GSi are trying to keep abreast of this, which has inevitably led to redundancies – 25 staff were made redundant in December 1994 and it is likely that there will be the same again in 1995. These are all from a department of 90 people and the process has occurred through voluntary severance. The staff knew it was coming but it did not make it any less difficult when it happened. The Operations Manager handled this very well. Some people see it as related to the TQ process but it was not related at all. TQ has not lost anybody a job and it is not the company's intention that it will. TQ should lead to increased business and ultimately greater job security.

The volume of business is growing. The company has a virtual monopoly in the marketplace and the trend, apart from the period during the Gulf War, has been upwards for many years.

The Data Preparation Manager:

Since we began the TQ process there has been a drop in the number of transactions that we are handling. I was looking after six external bureaux, now only one. As a consequence there have been some redundancies. I'm not sure how that is seen – whethere or not it's due to TQ – I really don't know.

Key stages in the implementation process – an overview

2.1 PHASE ONE – GETTING STARTED

2.1.1 The International Managers Conference in Malaga, Spain – 26 September 1990

It all began in August 1990 with a telephone call from the Divisional Managing Director (DMD) of GSi Travel and Transportation to Professor John Pike, Director of the Total Quality and Innovation Management Centre at Danbury (UK). The DMD wanted someone to make a presentation on Total Quality Management to their International Managers Conference in Ibiza (later changed to Malaga). Arrangements were made for Richard Barnes (RB) to meet the DMD and the UK Director (UKD) of GSi Travel and Transportation (UK) Limited, at the Park Lane Hotel, Piccadilly, London, on 23 August 1990.

Subsequently RB presented a one-day 'Introduction to Total Quality Management' on 26 September 1990, the objectives of which were stated as:

- To assist senior managers to gain an appropriate perspective of Total Quality Management and to understand its key concepts and how they have developed
- To enable participants to appreciate the primary role of quality in the strategic objectives of GSi Travel and Transportation
- To demonstrate the importance of the corporate image and the customer's perception of quality to retaining and building on the business
- To assist participants to understand how a total quality process may be implemented and the benefits of such a strategy
- To acquaint members of the management team with international successes deriving from the application of TQM

- To highlight the role of the individual in the successful implementation of Total Quality.

This programme was well enough received for RB to be invited to prepare a follow-on seminar, the details of which were discussed at a meeting on 1 November 1990.

2.1.2 The Clivedon follow-on conference - 13 December 1990

This follow-on seminar took place at Clivedon on 13 December 1990 basically for the heads and seconds in command of GSi Travel and Transportation from each European country in which they had a presence.
RB recalls:

> This again went well and I was asked to consider a third session in 1991 with a view to getting their heads to get TQM implementation started in all the relevant countries.
> A meeting took place between Philippe Leseuer and myself on 31 May 1991 to plan this day at their next international managers' conference, which was postponed a couple of times but eventually took place at Egham on 17 December 1991. The theme this time was tools and techniques and was probably the best of the three.

2.1.3 The second GSi International Managers Conference - Egham, 17 December 1991

Objectives

- To enable managers to distinguish between assignable and common causes of variation and the action to be taken in each case
- To enable participants to become familiar with the use of measurement techniques as an aid to continuous quality improvement
- To review the main methods of quality costing and apply them to a specific case familiar to participants
- To review the obstacles to the introduction of measurement and plan to avoid them
- To help participants appreciate the role of teamwork in problem-solving and the action needed to encourage a climate in which teamwork flourishes
- To assist participants to explore the possible uses of quality improvement tools in their own work situations

2.1.4 GSi Managers Conference in France - late June 1992

Following this workshop a fourth event took place in France towards the end of June 1992 but Richard Barnes was unable to attend due to other commitments on the dates in question.

2.1.5 TQM implementation proposals from Danbury TQM Centre, 24 November 1992

Following the conference in France, according to Richard:

> Contacts by me with the UKD and the DMD eventually led to a proposal to support TQM implementation in the UK (the company felt it was better for implementation if they used local consultants in each country at that time. This was later overtaken by events when Corporate GSi in France decided that all its divisions (T & T included) should use MSI (Management Systems International). The UKD and the DMD, however, stuck out for using us in the UK and got agreement for the strategy side with MSI providing team training in improvement tools and techniques.

The implementation proposals are set out below.

2.1.6 Total Quality Management strategy workshop, 17-18 December 1992

Acceptance of these proposals led to the preparation and conduct of a Total Quality Management Strategy Workshop held for the seven members of the Steering Group on 17-18 December 1992 at Crowthorne, Berkshire. The objectives for this workshop are set out below.

Objectives

- To assist the senior management of GSi Travel and Transportation (UK) to develop an appropriate perspective on Total Quality Management, through the examination of the concepts and philosophy involved and examples of best practice
- To assist participants to understand the difference between Quality Control, Quality Assurance and Quality Management Systems
- To establish a clear Business Mission for the business and define the Critical Success Factors (CSFs) for its achievement
- To identify the key measures for organizational performance related to the critical success factors
- To enable the management team to evaluate a range of different approaches to implementation, e.g. Crosby, Deming, Juran, and

determine a best-fit solution to meet the culture and mission of the organization and its critical success factors

- To enshrine the company's commitment to quality in a formal Quality Policy and outline a communication plan for its implementation
- To explore and develop a statement of organizational values or philosophy, based on the existing values of GSi
- To develop a shadow structure to enable the concepts and techniques to be put into practice, including the establishment of a Steering Committee and assignment of responsibilities and action plans for the principal processes involved in implementation
- To introduce participants to the basic tools and techniques for quality improvement aimed at maximizing employee involvement and encouraging teamwork and problem-solving

Workshop outputs

The outputs of this workshop were:

- Steering Group established
- Management culture survey conducted among top team
- Health check carried out on organizational systems, structure and style of operation.
- Mission and critical success factors identified
- Responsibilities for driving elements of the culture change process defined and assigned to members of the Steering Group

Recollections of members of the Steering Group/ Management Team of the implementation process

Getting started

The Managing Director:

It was a top down approach – I can't remember how we started. I came into the company with a strong view that we should be doing TQ (I was influenced by Tom Peters). In my previous company we tried to do it, but the rest of the directors were not interested. As I joined in December 1989 the contract was due for renewal in June 1990. In April 1990 the contract was in danger of going elsewhere. Philippe and I were very much in line that TQ was the way forward. Philippe wanted to do something in a European-wide context and we set up a one-day session at the international managers' conference in Malaga run by Richard

Barnes. We all got on well with Richard. Meanwhile, unbeknown to us, our parent company was investigating TQ and later decided to do it. The two worked quite well. By then we were moving ahead. Philippe said let's try it in the UK. Following this there was a series of other presentations leading up to the final launch in December 1992.

At the time we were working with another training company and had identified that our culture needed some preliminary work. We held a series of seminars in which everybody participated. Basically this was a team-building exercise to set the ground work. It consisted of a series of two-day sessions for all employees. These were about understanding your part in the company, which at that time was very compartmentalized. For example, the evening shift never met the day shift. People on the evening shift said 'we thought we were badly treated, now we realize everyone is badly treated.' From these sessions we developed a list of items – what they got out of the programme and what problems they had. We produced a list of issues to be addressed but we had no formal framework to address them.

The Operations Manager:

I started with the company in August 1992 and attended the course for members of the Steering Group in December 1992. There was a lot of talk about TQM going on in the background when I joined. I was recruited partly because I had an understanding of TQ from my previous company. I felt it was exciting to be involved from the start. The idea was sold to all of us by the MD and the Group MD. It was a top down process with complete support from the top. The parent company influence was quite strong. In terms of a pilot GSI (UK) were it – we were the guinea pigs.

The Data Preparation Manager:

Total Quality was mentioned over a number of months before the workshop with Richard Barnes (December 1992). I didn't understand the concepts but some people had an awareness of it. At the time I had just been promoted and I didn't really appreciate how important my role was going to be.

The Managing Director was the main driver of the process.

The Manager –Direct Marketing:

I first heard about the TQ process through the MD. He was heavily promoting Anglia Business School at the time. It also came through my line manager. My first exposure was in December 1992 but I had heard about it before that. I remember the Managing Director was the beacon.

2.1.7 The first company-wide culture survey – January 1993

During January 1993 RB conducted a company-wide culture survey amongst all employees of GSi Travel and Transportation (UK), and subsequently produced a report for consideration of action by the Steering Group.

The survey instrument (questionnaire) was designed by Richard Barnes, Principal Associate consultant, Total Quality and Innovation Management Centre. Small modifications were agreed with the client prior to its use. The work of Rensis Likert is acknowledged by the survey designer in providing the academic base for the survey instrument structure.

The survey consisted of 21 parameters which were expressed on the survey instrument as a continuum of behaviours reflecting the presence or absence of an approach to work and organization consistent with a Total Quality Management philosophy. Survey participants were asked to give their views on behaviour, attitudes and organization in the company as a whole, rather than their own personal approach to the issue.

Extracts from the survey report are given below.

Summary of conclusions

The overall culture is quite compatible with the introduction of TQM; however, there are some important elements which will have to be addressed if the process is to succeed.

The parameters, analysed on a causality basis, indicate that where there are barriers to TQM embedded in the culture, these arise from systems and structural issues such as performance management and communications, together with some relationships between departments. Parameters relating to management behaviour and employee attitudes in general scored well.

The negative aspects of culture are concentrated in areas such as communication, decision-making and recognition. These are most likely to affect the motivation and satisfaction of people within the organization but may also have a negative impact on customers despite good intentions and attitudes. They are also likely to restrict the empowerment principles of TQM.

The communications with, and involvement of, the technical support staff in the decision-making process needs to be urgently addressed.

Implications for TQM

Inhibitors. The perceived poor communication and decision-making processes may inhibit employee involvement – a key area in Total Quality – as they stand. The limited use made by management of recognition will also have a negative effect on this.

Drivers. Many companies find it difficult to introduce customer-oriented improvements in the early stages of TQM, instead having to concentrate on more internally focused issues such as waste reduction and inter-function relations in order to retain employee interest and enthusiasm. The culture at Camberley should allow the focus on the customer, which is at the heart of TQM, to be more easily adopted.

The prevailing management style provides a good foundation for TQM, but this needs to be supplemented by appropriate systems to encourage communication and involvement.

The perceived attitude toward quality, which has received the highest number of extreme value scores of any of these surveys to date, is very encouraging and should enable the company to get its message across effectively. This will aid employee involvement, given the opportunity being made available.

Key results. The openness of the decision-making process (Q7) scored lowest, whilst the following items were all rated as less than 3.5, indicating that:

- Communications are principally downwards with little opportunity to convey views to management or between functions (Q6)
- Recognition is unusual (Q15)
- Communications on business developments is poor (Q4)

The low scores shown above are reinforced by the observed clustering of extreme scores. The same three questions elicited scores of 0 or 1 in over 20% of the responses of the parameter.

The highest scores (over 5) occurred on customer orientation (Q1) and the prevailing approach to quality (Q3). The latter was scored at 7 or 8 by 75 (over 40%) of all the respondents.

There were no important differences between sub-groups when divided by age, length of service, position or numbers of previous employers (see comment above).

A number of different perceptions are apparent when the groups are divided by type of contract. The most pronounced differences usually involve low scores by the fixed-term contract group but since these make up only four of the returns, the results can be discounted. Leadership visibility (Q19) was scored particularly low by the casual contract group, but rather higher than the overall average by the part-time permanent group.

The most pronounced and regular differences appear when the averages are analysed at the department or function level. The lowest scores most often come from the technical support group, whilst there is a tendency to record somewhat more generous scores among the data capture group. Much lower than average scores were allocated by the technical support group to the following issues:

- Attitudes to customers and suppliers (Q1 and Q2)
- Attitude to quality (Q3)
- Objective setting (Q9)
- Use of management time (Q10)
- Means of employee motivation (Q13)

2.1.9 GSi Travel and transportation (UK) Limited Steering Group workshop – 25–26 January 1993

After the initial strategy workshop the Steering Group met again for another workshop on 25–26 January 1993. The objectives of this were as follows:

- To review and approve the terms of reference for the Steering Committee, agree the roles to be adopted and assist the members to develop action plans for the implementation of TQM within the company
- To review the results of the culture survey and the previous work on health checks and critical success factors in order to ensure the structure of the Steering Committee, its assigned roles and the key actions arising are consistent with the principal change areas identified
- To continue the process of deploying the Company Mission within the business through the use of objectives based on the critical success factors identified
- To introduce participants to a range of tools and techniques for quality improvement
- To identify potential fast-track projects for the development of early opportunities to apply the processes and tools of Total Quality
- To identify and approve the early steps to be taken on the communication and training plan for all employees

Workshop output

- Fast-track projects identified
- Initial stages of the communication and training plans developed
- Initial plans for culture change developed by individual members of the Steering Group

During the workshop participants completed an organizational health check and the following key areas arising from it were identified for further discussion:

1. Downward communication
2. Understanding customer requirements
3. Goal-setting and appraisal

4. Internal customer concept
5. Management training
6. Staff training in key skills
7. Consistency of messages in training

RECOLLECTIONS OF MEMBERS OF THE STEERING GROUP/MANAGEMENT TEAM OF THE IMPLEMENTATION PROCESS

Downward communications

The Managing Director:

This issue came out in the previous workshops. People didn't know what was going on. The communication side of things, we recognized, was something we had to improve. It was one of the critical areas – until then it was not a particularly open company. It was the philosophy of the group, but it was not practised much.

The TQM Training Manager:

The team briefing process which was already in place was strengthened so that managers were encouraged to disseminate information, not only about the TQ process, but about the business in general, all the way down through the organization. The group also planned for the introduction of a TQ newsletter which appeared in the summer of 1993.

The results of these improvements are that more questions are being asked about company matters and employees feel that they can ask questions more often than they could before. A side-effect of this process has been that people are asking more questions about what other departments are doing. This has facilitated the breaking down of barriers between departments and has improved cooperation between them. Whilst it is not possible to put a measurable value on this, there appears to be much less dissatisfaction in most areas.

The Manager – Direct Marketing:

After the Steering Group training there was a series of meetings to disseminate information about TQ. Team briefing was already in existence but these were irregular and had no fixed agenda. It was for immediate subordinates. Subsequently we became more focused in communicating at every level with regular, structured meetings. These were used for launching the TQ programme to all staff and to explain the aspirations for the company and its importance to us. The launch was contentious. Linda and Philippe disagreed. Linda wanted a big bang, Philippe didn't. He wanted it relatively informal. In hindsight we did the right thing. We had a phased launch. I can't remember exactly how we did it, only that there was a mixture of opinion.

The Operations Manager:
The concept of health checks doesn't stay with me. I was probably too new to the organization.

The Data Preparation Manager:
I never felt not communicated to by my bosses but get messages that there is a lack of communication. I'm not quite certain what it is they are looking for. There is a gap between the day shift and the evening shift. With the redundancies we have been careful to tell people what is happening. People do pick things up differently.

Understanding customer requirements

The Managing Director:
I felt we were very receptive. We never asked customers what they wanted in any formal way. Our main business started in 1984. The Direct Marketing side knew what their customers wanted. They already had a partnership approach. Another concern, apart from very few people understanding our customers' requirements, was that most people simply did what they were told without understanding what the results of their labour was.

The TQM Training Manager:
The health check increased the realization that there was a need to get closer to customers in order to obtain more information about their requirements. As a result a new full-time position was created, i.e. the Accounts Manager, which superseded the Product Development Manager position. The present incumbent of this position volunteered to take it on. In addition to the creation of this post, greater emphasis was placed on using other opportunities to find out what customers were thinking, via the considerable number of staff who are out and about.

The Data Preparation Manager:
After the initial training course there was a lot more emphasis on customer needs. A new position of Accounts Manager was created

in January 1994. Recently another Accounts Manager has been appointed to look after the BSP contract. Other parts of the company already had Account Managers so it was not a completely new idea. My main contact outside was with one of the airlines – so things didn't change much there.

The Manager – Direct Marketing

After the Steering Group training we set about finding out what customers wanted in service levels and products. We looked carefully at our customers and put a lot of time into it. We were afraid of raising expectations. We decided that we should not talk to customers until we could do something about it. As a result we didn't do anything. We are now just starting to do a survey but we already know our best customers anyway.

Goal-setting and appraisal

The Managing Director:

We didn't do it.

The TQM Training Manager:

Initially, following the health check, the Senior Management Team went through a goal-setting process at the CSF level and for a while there was some monitoring of accomplishments against these goals – but it did not last.

The Manager – Direct Marketing:

We took a long time to get started on this. We are starting appraisal now, at the beginning of 1995. QIDW has helped the introduction of appraisal.

Internal customer concept

The Managing Director:

I knew about this but I can't remember what happened. We were talking a lot about prevention. It was a big cultural change, especially for senior management to recognize that finding faults just before they went out the door was not the most effective way of practising Total Quality.

The Data Preparation Manager:

There were some misinterpretations of internal customer requirements. We still lose sight of who is the customer and who is the supplier – it's very confusing.

The Manager – Direct Marketing:
Yes, this is there.

Management training

The Managing Director:
We hadn't done a lot. We needed skills we didn't have. We needed a process for the management team since it was going to be the driving force for the whole programme.

The TQM Training Manager:
The Steering Group took a conscious decision to train all staff in the MSI programme and the first course was run in February 1993. This training was provided for 100 staff and consisted of two x three-day programmes (600 man days of training). It reinforced the notion of the internal customer concept and this in turn led to a reduction in the levels of conflict between departments. New staff ask their internal customers what they want. This happens all the time. It has also led to a reduction in the amount of rework.

A good example of this can be seen in the way the systems team now operates. They find out what their customers want before designing solutions which have to be reworked. It is not possible to put figures on these benefits although it would be useful if such a figure was available.

The training from Danbury and the MSI training integrated very well. There were some terminology differences such as Danbury's CSFs and MSI's Policy Management, but this was not a problem. Both approaches were compatible and reinforced each other. Throughout all the training sessions there appears to have been no resistance to these sessions.

Staff training in key skills

The Managing Director:
There were certain things needed in Total Quality. Staff training was not seen as a normal part of life except in the technical training area. The rest of the company was ignored. We went through a series of brainstorming sessions to try to get to grips with the situation. Our difficulty at that time was trying to overlay practical applications of TQ with changing the company's culture. The cultural roles were more subtle – they were not so touching-feeling things. This is where we had the most difficulties. We had to change at the top but it took us a long time to recognize it.

The Manager - Director Marketing:

We concentrated on MSI problem-solving and IT training. There's still a lack of training in other skills - we've forgotten about other training.

The Operations Manager:

With being a trainer, I told my people what to expect and where they were going. The briefings given depended upon different departments. I never got round to formalizing what the corporate role was although there was an evaluation process. A lot of people were frightened of mixing with people they didn't know.

There was a fear of people even within GSi - anybody not in the same room were regarded as strangers. There was also concern about the hierarchy. Some people missed attending courses with their peer groups and this led to even more fear.

The Data Preparation Manager:

Within the department we have always ensured that people promoted have always been trained in the technical skills they require so they understand the systems we use. Two supervisors in the team received sponsor training with MSI. We have a couple of groups in my own area. We have three pilot projects and I'm involved in one of them. The initial project selection seemed like a good idea but projects went on for a long time. We got entrenched in them, but they were very useful as a learning experience.

Consistency of messages in training

The Manager - Direct Marketing:

There is a consistency in the messages coming from the training programmes. IT training is seen as an integral part of the quality programme.

The Operations Manager:

There was a conscious knitting together of the two approaches from Danbury and MSI. Danbury helped to establish the structure and the implementation process. MSI's Quality in Daily Work process armed people with the tools and techniques to do it. If you like, Danbury created the building, MSI put in the furniture.

Critical success factors

Also during the workshop the Steering Group were asked to identify and agree the most critical success factors for their business. The following seven CSFs were agreed:

- Develop our relationship with individual customers to secure our market position
- Provide reliable systems and procedures
- Manage suppliers to achieve 100% conformance to specification
- Remove expectation amongst staff and customers that new or changed procedures, methods or systems will inevitably contain defects
- Build confidence in our knowledge of their business
- Provide solutions which customers buy
- Develop all staff to achieve their maximum potential

Recollections of members of the Steering Group/ Management Team of the implementation process

Critical success factors

The Managing Director:

Basically these were brainstormed at the Steering Group workshop in December 1992. There was some debate about some but not about others. They gave us a bit of a focus and we felt they should affect what we were doing. They gave us an opportunity to set targets and items for measurement. They were very important at the time and were critical in explaining to the rest of the company why we were doing what we were doing. Some of the CSFs were quite difficult to measure. After two years we decided to reassess them. Richard Barnes said at the time that we would. Now we have moved to policies. We are saying the same things using different terminology. We have not changed our view about any one of these. We have simply reworded them more effectively.

The Data Preparation Manager:

At the time when we started I lacked confidence in a group situation. I appreciated the difficulties involved but found the whole thing very daunting. We were trying to develop our own staff to take things off us – delegating things downwards. It seemed that there were too many CSFs. They seemed valid at the time but we wondered how valid they were if we were not measuring them.

Now we have four out of the original seven. Instead of having a specific responsibility for one of these, now no one takes specific responsibility for individual items. They all feed into all of the decisions.

The Operations Manager:

I didn't like these at first – I couldn't say it – I used to say Critical Sex Factors! I prefer the term 'policies'. I realize how difficult it was to measure the CSFs. The change of name was not significant. We have become more concerned with things we can measure. At the beginning we were a bit ambitious, we started with seven. Now we are down to four. The initial seven were based on our culture roles and were intended to provide a measure for each role.

The Accounts Manager – BA Executive Club and BSP:

We have seven CSFs, one for every member of the Steering Group. Projects always linked into customer satisfaction. At that stage all that was important was to get as many people involved. The formal linking in of these came later. The question was asked at the Steering Group. We had Quality Circles before we were doing Departmental Purpose Analysis. We were lucky we didn't fall down too many holes. Then DPA led to setting up new Quality Circles in each area.

Identification of cross-functional 'fast-track' projects

A major outcome from this Steering Group workshop was the decision to establish a number of cross-functional fast-track projects. The following projects were agreed:

- Net remit – team leader – the Operations Manager
- Method and processing – the TQM Training Manager
- Invoicing – the Accounts Manager

These were subsequently confirmed at the Steering Group meeting on 26 February 1993.

Allocation of culture change roles

Finally, from this workshop came the allocation of cultural change roles. These were assigned as follows:

Position	**Role element**
Managing Director	Management Obsession and Measurement
TQM Training Manager	Costs of Quality and Corrective Action System
Data Preparation Manager	Supplier Conformity and Customer Care
Operations Manager	Training and Education
Direct Marketing Manager	Reward and Recognition
Major Accounts Manager	Communications
Accounts Manager	Employee Participation

Recollections of members of the Steering Group/ Management Team of the implementation process

Application of culture change roles

The Managing Director – Management Obsession and Measurement:

I was the beacon. My role was Management Obsession and Measurement. Linda was Communication; Jane was Employee Involvement; Phil Russell's was Reward and Recognition; Alistair was Cost of Quality and Corrective Action System; Lesley Goodall was Training (she fought for it) and Lesley Holmes was Customers and Suppliers.

'Obsession' was a bit easier than the 'measurement' bit. Shortly after the December Steering Group workshop there was an opportunity for me to go to a workshop run by MSI whom the group had brought in. This was on tools and techniques for problem-solving. I was so enthusiastic about the approach, I was committed to get Alistair and Lesley Goodall trained in order to set up our own training service so we could take all these changes into all departments. I tried to become personally identified with the training programme by talking to groups of 25 people at a time. I put the training in the context of the business and what it meant for us. I also ran two training sessions for supervisors with Richard Barnes. In general I tried to keep it at the top of the agenda. I have been critical of people in the past, but I tried to be error friendly, focusing more on the causes of problems and how to stop them happening again.

Measurement:

There was clearly a need for it but it has taken us a long time to get anything in place. We have not always been successful.

It has only been in the last six months, to be honest, that we have got to a position where measurement is seen as a tool set and can tell us an enormous amount about how we are doing. This is due to other people recognizing its strengths. Looking back, the most difficult part for people has been in deciding what to measure.

The Data Preparation Manager - Customer Satisfaction and Supplier Participation:

My role was Customer Satisfaction and Supplier Participation. I haven't really done anything in that role. It's been in the background, but I had a lot of other issues to deal with and was kept busy trying to do the day-to-day running of the department. I didn't relate TQ to the business at that time. As you get into it you begin to see that it is to do with the running of the business. The barriers for me were personal ones. Since the Steering Group was disbanded, Total Quality is seen as running the business - doing the whole thing together.

The Operations Manager - Training and Education:

The symbol for my role was the tree. I volunteered for Training and Development. I went on to become one of the MSI trainers. I'm still doing it but am not very active because of the pressures of my operational role. As I have developed in the role, I've not had so much time. I did put forward a number of suggestions to the Steering Group. This is a wonderful company to work for in that there are no strict procedures that everybody has to follow. I introduced a training needs analysis and a training plan within my own area of responsibility which was very successful. It has enabled us to cope with making 25 people redundant. We needed staff to be more flexible, which they are now. The staff can see the redundancies are a result of technology, not the TQ process.

The Manager - Direct Marketing - Reward and Recognition:

My role was reward and recognition. I looked at reward and recognition schemes run by the company and attempted to benchmark these against other companies like BT. I also looked at appraisal systems and got examples from other organizations like the local TEC and BT again. I used BT as a model, as a starting point. It was complicated at the time. The key was measurement but BT didn't lend itself to measurement. Apart from appraisal, not a lot has happened. The main reason for this was a lack of general interest in the Steering Group. It was an emotive issue and difficult to get consensus. Historically each department rewarded staff in different ways – for example through productivity bonuses, £50 vouchers at Christmas and so on. The priorities lay in other areas of the business. I, personally, was not totally committed to TQ. I think a couple of the managers used their change roles to further their own careers and aspirations. Interests outside the company drove them forward.

The Accounts Manager – BA Executive Club and BSP – Employee Participation:

My role was Employee Participation. I was taken off the Steering Group some time back. Since then I've had little involvement except for TQ being part of my daily work. My role lasted for about twelve to eighteen months actively. I don't know now whether I held it or not. Initially I was responsible for getting Quality Circles off the ground – establishing their terms of reference. The first thing was setting up the Shadow Structure – for example whether the Steering Group was managing the Quality Circles or whether it was the line managers.

It also included how people should manage meetings – keeping time. The MSI courses taught us to set objectives, prepare a plan and manage time for meetings (OPT). It was a neat little acronym. In carrying out my role I would go out and give Quality Circles, twenty minutes informal presentations. Get them to identify who their sponsor was, and ask them 'Are you sure you've got your project right?' It was decided that the Quality Circles should report to line managers, whereas the Cross-Functional Fast-Track Projects have a sponsor who initially reported to the Steering Group which is now the Senior Management Team.

I was involved in the original courses. PR, the head of Direct Marketing, asked me to get involved. PL and GS were the main champions. There was a course in France about Getting to Know GSi and the big plans for TQ. I was aware that it was company wide. In the UK we were taking a slightly different approach.

Other culture change roles

Other culture change roles were not explored since there were no interviews scheduled with the managers concerned.

The TQM Training Manager's view – Management Obsession:

This has been there all the time. It is noticeable within the management team level. The Managing Director talks quality all the time.

Now there is a strong focus on target setting and measurement against targets which has emerged over the last year. The style of the management team has changed (thankfully). It went through a stage of trying to get consensus but got nothing done. Now the Managing Director is more proactive. People are saying they enjoy meetings more because more gets done.

Customer Satisfaction and Supplier Participation:
Nothing.

Training and Education:
The company tackled training and came up with a training plan to which they adhered. Project Teams were established and trained in tools and techniques or improvement technology in MSI terms. However, the company did not train managers in project selection and consequently paid the penalty for not looking closely at what people wanted to do. To date (February 1995) the company has completed about twelve projects. Towards the end of the summer of 1994 a conscious decision was taken to freeze projects until managers had clarified their thoughts about the relationship between CSF and policy management.

Reward and Recognition:
Nothing.

Communication:
This has produced a lot of success.

Employee Participation:
The improvement team worked on guidelines and procedures for setting up the team, but these were not adhered to mainly because the corporate culture is not one of following written procedures.

The Operation Manager's view:
Phil Russell looked after Rewards and Recognition. He did a lot of work on this. He's still actively looking at it. As a company, we seem to be very good at debating things but not very good at implementing things.

Jane looked after Shadow Structure. She developed mechanisms for the Quality Circles to operate. Linda's role was Communications. She introduced a *Quality Newsletter* which has been very popular. Greg was the beacon – that is Management Obsession but the role passed to Alistair. Otherwise . . .?

The Manager – Direct Marketing's view:

We successfully introduced Improvement Technology training and the results were very successful. Communication was, again, successful. For example, we launched a TQ magazine and have produced four to five editions (see culture survey reports).

2.2 PHASE TWO – AWARENESS AND 'FAST-TRACK' PROJECT TEAM TRAINING

2.2.1 First meeting of Steering Group – 8 February 1993

The outputs from the above Steering Group workshop brought Phase 1 to a conclusion. Phase 2 began with the first meeting of the Steering Group. The main emphasis of this phase was originally set out in the Danbury TQM Centre's Implementation Proposals. However, at this first Steering Group meeting a decision was taken to use a training company called MSI (Management Systems International) formed from managers involved in an American company called Florida Power & Light which had won the Deming prize. MSI were contracted to provide training in other parts of the GSi Group and it was agreed that their approach would be integrated with the Danbury approach. MSI provided training in tools and techniques for problem-solving and a series of templates (1–6) for use by project teams and Quality Circles. MSI also provided the training for the project team leaders and sponsors.

Other issues discussed at this first meeting were:

- Training plans for SG members to attend MSI Team Leader and Train the Trainer courses to assess their suitability for GSi Travel and Transportation UK
- Communication of culture survey results
- Team briefings to recommence

Staff recollections of how they came to know about TQM

My first recollections of TQM were from the MSI courses. Part One of the course was held in February, May and September 1993. I went in February and then about May or June. they were three days each. Then I started in a group.

'We had mixed feelings. I thought – great – at last they are paying attention to staff. But I had heard about TQM from other companies through friends from outside or from staff in GSi and went "oh no, not one of these projects".'

'For me it was necessary as I was a member of a group and we didn't really know what we were doing.'

'I didn't know anybody involved so I didn't know anything about it.'

'At first I was a bit confused. Then I went on the course and became even more confused. It didn't become clearer until the end of Part Two. When I heard about it I was nervous and apprehensive about what was going to be expected from me. They should have clarified at the beginning of the course what it was all about. It was more them talking to us, rather than giving us a practical understanding.'

'I was introduced to TQ two years ago.'

'I asked to be put on a team and was told I would need to learn the techniques. I went on the MSI course to learn about the templates then I came back and got involved in a team. I went on the second course a week before my wedding so I had a lot of things on my mind. At the same time I went on the course on DPA with Richard Barnes.'

'We were all sent on the MSI training programme. We went in the second lot.'

'We were very nervous because we had never done anything like it before. It was quite hard – Pareto and all that.'

'The gap between courses was very wide – it was really too long. A lot of what we learned was useful. It was an eye opener. We had no use for the tools straight away so we got out of the habit of how to use them. We don't always use the tools in our everyday job.'

'They could have been more selective about who they sent. People have been on training but have not done anything with the training so the knowledge is wasted. You need to get into a team straight away otherwise you forget. Managers "pass the buck" when things go wrong.'

'There shouldn't have been such a big gap between the two parts of the course. There was also too much to take in in one go. Some people were completely turned off by TQM on these courses. It seemed complicated, it was a strange environment and people came from different countries.'

'There was another instance though of someone who started off very anti, who came round, and another who was lacking confidence yet enjoyed it.'

2.2.2 First 'fast-track' project meeting – 15 March 1993

The first 'fast-track' meeting was held for the Method One Processing project.

2.2.3 Sixth Steering Group meeting – 25 March 1993

This was a special Steering Group meeting with RB to sort out confusion and concerns over the role of the consultant and the individual culture change responsibilities, including how each role interacts with each other. Much improved clarity on individual roles was established.

2.2.4 First TQM newsletter issued – April 1993

See section A1.3 in the appendix to this case study for extracts from a typical newsletter.

2.2.5 Presentations by UKD and Management Team – April 1993

Presentations were given by the UKD and members of the management team to all staff on TQM and why it was important to the company.

2.2.6 Team building interviews and workshop – 26–27 May 1993

Team building interviews and a workshop was held for Steering Group members with Team Development Group (external consultants).

Recollections of members of the Steering Group/ Management Team of the implementation process

Team building

The TQM Training Manager:

The effects of the team building exercise were short term. Members of the Senior Management Team became more aware of each other's needs but it did not last long. For a while they were more willing to concede to the needs of others. The problems that were there before are still there. As a result of the TQ process the Senior Management Team met more regularly and talked about issues, but did not always make decisions. Often decisions are taken outside of the meetings. With the passage of time and new blood, this has added a new dimension. With the dissolution of the Steering Group there have been two additions and one departure – 'we need to see how it works out'. The targets now set act as a driver for enhanced cooperation.

The Operations Manager:
The team building exercise created an awareness of things that wouldn't have come out in ordinary circumstances. It helped us to identify each other's strengths and weaknesses. I think we could have built on this – it was a missed opportunity. Some did it. It was very emotional for some people for sure. The thing that stayed with me was that it all centred around 'fathers'.

The Manager – Direct Marketing:
The team dynamics have improved for a couple of reasons. The first is, membership of the team has changed. The second is that we are all more knowledgeable and more focused about what we are trying to achieve. The effects of the team building exercise were short term. At the time we discovered more about each other – about our weaknesses and strengths and we learned to be more open with each other. But we go through phases – sometimes we are open, then we all clam up and say nothing. Other things are more important such as membership and focus. We now combine TQM and the business. It was a big mistake looking at TQM in isolation from the business. TQ is an integral part of the business strategy.

The Data Preparation Manager:
We now feel we are definitely working more closely as a team following the team building course in May 1993. This had a long-term benefit on the Senior Management Team. The team dynamics, however, has changed because it has got two new members and one member no longer belongs. Additional training wouldn't be a bad thing, but I personally don't feel any concern one way or another.

Originally team dynamics was a problem. As well as a lack of experience, we found it difficult to focus. At the start I didn't know enough to be able to contribute. Sometimes some people have problems identifying their value to the team, especially when working with other departments. Now people are getting used to working in a team. Normally we are isolated, working in our own little departments. When people from other departments didn't understand what I was trying to say, I tended to shout at them – you know, like when you can't speak a foreign language you shout. All suffered from this. I was impatient. Now we are learning to communicate with each other.

The Accounts Manager – BA Executive Club and BSP:

We also concentrated on team building skills and how to get consensus from the Steering Group. We looked at the dynamics of the group and we were aware that it was not working very well. MR was brought in to run the team building course. We used to vary from week to week, sometimes we were very productive we would stick to the points and felt we had achieved things. Other times we would get stuck on one subject and we would get away from the agenda. We were not plugged down to one thing. People were trying to achieve things but couldn't really get there. We were unhappy about handing over things to other people so the departments had to give leadership to the troops. But we needed to be doing something about it ourselves.

In the departments we used MSI courses which brought out lots of points about what to look out for.

After the team building course we certainly got to know things about each other. I didn't come back and do things I was meant to do mainly because of pressure of work. You need to be encouraged in things like that. We were going through a difficult patch in the organization. The course had a lasting effect on me for some time. The sharing part people found uncomfortable. I look back on it as an uncomfortable experience. I don't think it was constructive but it wasn't a bad thing. Our group was not close enough to share. It didn't make me feel closer to others. On the first evening I tried to relax. I wanted to put the day behind me – it was too overwhelming, too strong, I couldn't handle it. I wasn't ready for it but I would do it again. In the right circumstances it would have worked.

2.2.7 The revised proposal for TQM implementation – 30 May 1993

Revised implementation proposals from Danbury TQM Centre were submitted following decision by GSi at corporate level to provide a common format for problem-solving and improvement projects based on a format developed by MSI (Management Systems International). The revised sections of the proposal are set out in the appendix to this case study (see section A1.2).

2.2.8 Acceptance of proposals

Again the proposals were accepted and the implementation process continued according to plan.

2.2.9 MSI team leader training – 7–9 June 1993

MSI team leader training workshops for fast-track project team leaders were held.

Recollections of members of the Steering Group/ Management Team of the implementation process

Cross-functional 'fast-track' projects

The Managing Director:

The cross-functional fast-track project teams were, in hindsight, given the wrong problems and they were not managed closely enough. They suffered from what most teams suffered from – that is, a lack of direction. Now we have found people capable of driving them forward. The projects were poorly chosen, their scope was not clear. The team members needed more coaching. They all achieved results, so that was good.

The TQM Training Manager:

Critical success factors were identified and three cross-functional fast-track project teams were established. Of approximately eighteen middle managers, four or five were nominated and invited to participate in these teams because of their expertise which was considered relevant to the projects. Initially there was some resistance to the suggestion, mainly from technical people who feared the additional workload involved. However, this initial resistance was overcome by convincing them that their time would be covered by the recruitment of additional casual staff. In Systems and Programming the problem was never really addressed. The Steering Group knew it was there – it was always an issue, but it did not find a solution.

Now this group is probably less involved. Nevertheless, the recently employed new manager will need to look at the resource issues for the future. Because the department employs a relatively high number of contract staff rather than permanent staff, it is not so easy to select permanent staff to work in improvement teams. The same people are always being asked to participate. This compounds the problem of work overload.

Middle Managers who were not initially asked to join the fast-track teams felt a sense of great relief that they were not being asked to take on extra work. Amongst the clerical staff who were asked to participate, most of them willingly joined in without any resistance.

Eight months after the conduct of the senior management strategic workshop, middle managers and key professional/technical staff participated in a Management Awareness Programme which was triggered by them because they wanted to know more about what the TQ programme was all about. The Steering Group regarded it as a communications exercise to spread the word and hopefully put into context what they were trying to do.

The Operations Manager:

I ran one of the fast-track projects – 'Net Remit' – which gave us the chance to put everything we were taught into operation. It took about a year and we learned a lot from it. It also helped with the teams that have formed since. We applied the learning to the Quality Circles. One of the most important things we learned was working together as a team. People were more visible in a small team. When you are training people it becomes very clear that people communicate with each other differently. It was one hell of a learning curve. We were too ambitious, the team was too big. To begin with we all got quite excited. Since we've stratified things, things went quickly after that.

The original team leaders were selected from the Senior Management Team. They went for training. We selected the problems first then the team leaders and the Senior Management Team identified the members of the cross-functional teams. The senior managers for each area approached the nominated members to ask them if they would like to join the team. There was some fear of being nominated but a lot of support was given to convince them. There were some fears about being in the same team as their boss. However, the interest in the problem itself overcame their initial fears. I bribed them with chocolate biscuits and orange juice.

The Data Preparation Manager:

One of the current projects is the Net Remit project. This is to do with agents and airlines. We had a lot of problems with what we call invalid rejections. We're trying to identify the causes of these in order to reduce them. We use the MSI templates to tackle the problem. These tools and techniques are very helpful. Full records of these projects are kept identifying how much time is spent on them etc.

The Accounts Manager - BA Executive Club and BSP:

There were three fast-track projects which were the slowest fast-tracks ever! In addition to this we had Departmental Improvement Groups (DIGs). Everybody was voluntary. We recruited the project leader - usually a supervisor or senior member of the team. We were not very good at setting project titles.

2.2.1 Training needs analysis - 17 July 1993

It will be noted from the calendar of events that the Steering Group decided at its meeting on 11 March 1993 to ask Richard Barnes to facilitate a training needs analysis.

The training needs interviews were carried out during June and July 1993 with all members of the Steering Group. The process involved a structured interview together with the use of a card set known as SPIO (Specific Problems, Issues and Opportunities) to bring out the major areas of concern over personal performance.

Following this a report was prepared which indicated the high priority training needs and type of training delivery process best suited to each person. It was issued on 17 July 1993. An appendix gave a breakdown of the key personal concerns reflected in the cards chosen from the set used during the interviews and concluded with a summary of general training needs which could be met within the team itself rather than individually. In particular, the acquisition of management skills appropriate to leadership in a TQM culture was emphasized.

Main recommendations for short- to medium-term actions arising from the training needs analysis

These were stated as:

Short term

1. Follow up this report with individual interviews to confirm/modify and act on the training needs identified here

2. Source training for team on coaching and counselling skills
3. Source further training for trainers on general trainer skills

Medium term

4. Identify a suitable technique for long-term change planning and arrange for training to support it
5. Source training in negotiation skills for those members of the team expressing concern in this area, particularly those who need it as a specific job skill

Recollections of members of the Steering Group/ Management Team of the implementation process

The TQM Training Manager:

A training needs analysis was conducted by Richard Barnes but the results were not used. There is still training going on in other skills. This is an ongoing process. For example, managers have been trained in team building skills to support the TQ initiative. There is a consistency in messages coming through from all the training.

The Data Preparation Manager:

Members of the Senior Management Team have just completed a training course - using external trainers via the corporate training department. A training needs analysis was done by Richard Barnes. Only part of that was appraisal training. I personally felt I could do with some management training. I did discuss the TNA with the MD but I haven't gone through the written report - it's still on the table - I'm seeing RB tomorrow.

2.2.11 First Quality Circle established - 3 August 1993

Recollections of members of the Steering Group/ Management Team of the implementation process

The Data Preparation Manager:

Currently we have one Quality Circle coming to an end. Now we are encouraging people to use the problem-solving templates in their daily work. The existing Quality Circle was dealing with 'data captures that reach the edits' – to stop rework. It's been running since July 1994 with a few stops and starts. The group has been making small changes as they have gone along. The problems have been reduced by 50%. The Data Preparation Manager is the sponsor for the team. At the MSI review it was thought that the team was doing well. Not only are we more focused, we are probably doing more now than before.

The Accounts Manager – BA Executive Club and BSP:
Setting up Quality Circles took a lot of time. They were not in every area until well after September 1993. Altogether we've had fifteen Quality Circles. AC has records. In the bigger departments there were perhaps three – to do with how we could manage new techniques. People weren't aware they could actually do something about problems and get their line manager or senior supervisor to help them out. We tried to ensure that every area had a QC working.

2.2.12 Departmental purpose analysis (DPA)

A pilot scheme for implementing departmental purpose analysis was started in the data preparation department ahead of the planned training programme.

2.2.13 TQM management awareness programme – August 1993

During June 1993 the Steering Group discussed the need for a TQM management awareness programme and requested Richard Barnes to prepare an appropriate proposal. Workshops were subsequently conducted on two occasions and involved substantial inputs to the training of members of the Steering Group and the preparation of an on-site training room in themes appropriate to the company's industry. The objectives for the workshops are set out below.

For whom the course is intended

Managers, supervisors and key professional staff who have not previously been involved in TQM concepts training.

Objectives

- To assist participants to develop an appropriate perspective on Total Quality Management, through an examination of the concepts and philosophy involved and examples of successful approaches

- To assist participants to understand the difference between Quality Control, Quality Assurance and Quality Management Systems
- To enable participants to differentiate between prevention, appraisal and failure costs and identify the relevance to quality costing as a tool for identifying improvement opportunities
- To encourage participants to recognize and use appropriate leadership behaviour needed to ensure the systematic solution of problems, prevention of defects and the involvement of the total workforce in the continuous improvement of products, services and processes
- To give groups of participants the opportunity to work through the internal customer/supplier chain which determines the quality of the end product and to select appropriate improvement targets
- To commence the deployment of the Company Mission throughout the business

2.3 PHASE THREE - MANAGING THE IMPROVEMENT PROCESS

2.3.1 Team sponsor training - 20–23 September 1993

This was conducted by MSI for members of the Steering Group who were sponsoring projects.

Recollections of members of the Steering Group/ Management Team of the implementation process

Departmental project teams

The Managing Director:
Some have been extremely effective, others have lost their way in the same way that the cross-functional teams did relative to the amount of coaching and sponsorship and direction they received. Most suffered because they didn't narrow the problems down enough. It was not the fault of the teams. It was a lack of coaching. Now the new teams are growing in confidence

The TQM Training Manager:
Over the two years from the start of the process up until the end of the summer of 1994, the company had completed about 12 projects. However, the Senior Management Team took a consensus decision to temporarily freeze projects until the issue of

critical success factors and their relationship to the MSI concept of Policy Management was clarified.

Now the company is starting three more projects. These are:

1. Direct marketing
2. Reducing idle time for staff working on the switchboard (estimates show that there are 500 hours of idle time per annum)
3. Reducing rework in correspondence.

2.3.2 Cost of quality workshop – 17 October 1993

During August 1993 the Steering Group discussed the need for a workshop on quality costing and at its meeting on 13 August the delegates were identified who were charged with investigating and designing a process for reporting on quality costs within the company. The course was scheduled for 17 October 1993 and the objectives for the programme were:

- To assist participants who have previously been through TQM awareness training to appreciate the role of quality costing as a tool for continuous improvement
- To introduce the cost of quality team at GSi to a range of techniques for identifying and reporting quality costs
- To consider a range of methods for displaying and using quality costs
- To establish actions and roles for the members of the cost of quality team at GSi, to support the collection of the first batch of quality costs

Following the workshop it was reported to the Steering Group at its meeting on 1 November 1993 that the cost of quality meetings were continuing and that participants now had a better understanding of the terminology. Plans for collecting and displaying costs of quality information were subsequently agreed.

Recollections of members of the Steering Group/ Management Team of the implementation process

Cost of quality and corrective action system

The TQM Training Manager:

A system was put in for collecting cost of quality information following a training course conducted on 17 October 1993. The information, however, was not used. It was collected and displayed but nobody did anything with it. Since December 1994 this has changed. The company has now started to use it.

2.3.3 First draft of detailed master plan – December 1993

The first draft of a detailed master plan for Steering Group culture change roles was produced.

2.3.4 First set of measures for critical success factors (CSFs) produced – 17 January 1994

The first set of measures for CSFs were agreed at the 22nd Steering Group meeting.

2.3.5 Presentation from cost of quality team – 31 January 1994

The 23rd Steering Group meeting received a presentation of initial findings from the cost of quality team and established further objectives for the exercise.

2.3.6 Departmental purpose analysis weaknesses identified – 31 January 1994

The Steering Group discussed the weaknesses of the DPA process. It was decided that DPAs should be progressed only to the stage of establishing the customer/supplier chain, but not the detailed customer requirements, until they had considered how DPA could be linked into an MSI framework known as Quality In Daily Work (QIDW).

Recollections of members of the Steering Group/Management Team of the implementation process

The TQM Training Manager:

> To begin with the company went straight into the adoption of departmental purpose analysis. This introduced the concept of the internal customer and staff became more aware of whose needs they were serving. However, following the training by MSI, the DPA process was quickly dropped and the MSI 'Quality in Daily Work' concept was pursued instead. This approach was felt to be simpler and more visual. It gives the company a clearer focus on customers and suppliers and provides information for determining improvements by setting and measuring indicators for key processes.

Staff talking about departmental purpose analysis (DPA) and quality in daily work (QIDW)

'We started DPA and got half way through it, then it finished. It was quite good. Then it just stopped and we changed to QIDW. No reasons were given. There was a gap between doing DPA and being told about QIDW.'

'I see them as quite different. DPA was quite good – it made you break down processes. We came up with a lot of good ideas of how to improve the system and I put things into the suggestion box.'

'With QIDW you are looking at yourself. One didn't cancel out the other. We can still use DPA, both are beneficial. In fact we do use QIDW, for example, in appraisals.'

'I'm more aware now of what I'm accountable for. It's promising what you know you can do. Everything I do – I'm measuring it. It sets targets for doing quality checks on the files. It acts as a memory jogger as well. Different shifts do different things. Where we don't do things wrong we don't measure it any more.'

'I ran the DPA team. I was promoted to supervisor, I didn't know how to lead it. It was all too much. We had five months to complete the task. The outcome from it was all presented in a folder but went on the shelf and was not used.'

'We did part of a DPA exercise about a year ago. We were quite enjoying it but it was stopped. Again we felt disappointed'.

'We did DPA to a limited extent in a different department. QIDW is different in some ways because I will be working with supervisors below me. I think it will be better. In DPA we were working in teams.'

'We began QIDW in October1994. They are putting the Credits Department through it.'

'There was myself and the supervisor and a team of nine people and we are taking it down to team member level. Each of us has to set out own accountabilities.'

'QIDW – I haven't been introduced to it yet.'

'With QIDW I know what my accountabilities are and now we are using the templates more and more especially since the new year.'

'In my area nobody was involved in projects but all of us are involved in QIDW.'

2.3.7 Merging the Steering Group with the Senior Management Team meeting – 14 March 1994

The decision to merge the Steering Group and the Management Team meeting took place on 14 March 1994 at the twenty-fifth meeting of the Steering Group. The new merged group met for the first time on 21 March 1994. This was done in order to fully integrate total Quality into the normal business management process.

Recollections of members of the Steering Group/ Management Team of the implementation process

Merging the Steering Group and the Management Team

The Managing Director:

We didn't really dissolve the Steering Group. We had a Management Team and a Steering Group. One of the difficulties was that they were seen as separate processes so we simply put them together. The agenda for the Senior Management Team is totally quality driven. We don't talk about the business, we talk about policies. We now spend more time on it. We are more conscious of the need to see it as the only way forward.

The TQM Training Manager:

Business strategy is now more focused on what the company needs to do over the next 12–36 months than ever before. Nowadays the view is that there is no difference between the business plan and the TQ process. Both are about building the business. Initially there was a need to have a separate Steering Group for the TQ process in order to ensure that it was given the visibility and priority that it deserved. Today there is no need to have a separate Steering Group since TQ is fully integrated with the business and TQ matters are indicated in the agenda of the normal Senior Management Team Meetings.

This has not meant a reduction in the priorities given to the improvement process. The knowledge and experience gained over the last two years has been invaluable. As a result, the company now has the structures and the mechanisms right to maintain the impetus. It has meant, however, that one member of the Steering Group, who was not a member of the Senior Management Team, now no longer participates in the Senior Management Team decision-making process.

The Manager - Direct Marketing:
There was no change in terms of the amount of time spent on quality when the Steering Group was integrated with the Management Team, and I don't think the rest of the company noticed any difference. Previously there was no link between the quality programme and the financial growth of the business. Quality is now in most aspects of the agenda at the moment.

2.3.8 Quality in Daily Work: accountabilities and processes - 14 March 1994

Dates agreed for completion and presentation of accountabilities and processes within the QIDW framework.

2.3.9 First merged Total Quality and Steering Group management meeting - 21 March 1994

2.3.10 Project Reviews - 6 April 1994

Project reviews were conducted by Professor Kano, a Japanese 'guru' appointed by GSi at corporate level.

2.3.11 Establishment of a Corrective Action Review Group (CARG) - 18 July 1994

The eighth 'merged' Management Meeting established a Corrective Action Review Group.

A new process for project selection, team management and measurement was also discussed.

TQ displays, QIDW roll-out plan, feedback on appraisal forms and appraisal training was also planned.

Recollections of members of the Steering Group/ Management Team of the implementation process

Corrective action system

The TQM Training Manager:
Mechanisms were put in place but the processes are not used.

2.3.12 TQ94 event - August 1994

A series of off-site presentations were made by the TQM Training Manager on Total Quality next steps and Quality in Daily Work (QIDW), and by the UKD on the future of the business. In between presentations staff had the opportunity to peruse the TQ projects and ask questions.

2.3.13 Review of critical success factors and adoption of policies - 5 December 1994

The fifteenth Management Meeting reviewed the lack of progress in the CSFs which were identified two years ago - several had never been measured and the value of those that had needed to be reviewed. It was proposed to adopt a new set of **policies** in place of CSFs in keeping with known aims. Appropriate measures were to be put in place to provide an appropriate level of information on which to base decisions. Targets needed to be set once measures were agreed.

An action plan was agreed as follows:

1. Key tasks to be established for each quarter in 1995
2. Monthly plans to be agreed
3. One policy to be reviewed at each meeting

Recollections of members of the Steering Group/ Management Team of the implementation process

CSFs versus policy management

The TQM Training Manager:

Following a review with MSI towards the end of 1994, it became clear that the company did not have sufficient information upon which to make decisions. Managers did not go through a process of target setting and had no clear objectives. The general excuse used was that people did not have enough time to analyse situations facing them. The review highlighted the need to rewrite what is now called policies and a paper on this was presented to the Steering Group

In December 1994 the concept of critical success factors (CSFs) was abandoned. When they were first introduced after the initial strategy training workshop in December 1992, they seemed like a good idea at the time and senior managers believed in them. Today they are no longer regarded as valid. The principles underlying CSFs *per se* have not gone away and the process of selecting them was not a waste of time. Senior managers feel they learned a lot from doing so. However, in hindsight, the range of CSFs was too vast. they were also too diverse and gave managers too much to focus on and too many things to tackle. The policy management concept which replaced CSFs enabled managers to formulate much clearer statements and caused them to focus down on four key issues rather than the initial seven CSFs. These four policy management statements are:

- Improving customer satisfaction
- Time to market
- Improving productivity and eliminating rework
- Increasing revenue

The process of arriving at these four entailed considerable healthy debate amongst the strong personalities who were members of the Senior Management Team. No one wanted to drop the CSFs which they thought important to them. However, at least one member of the Senior Management Team believes they should have been encouraged to focus down on only four CSFs from the beginning.

2.3.14 Second company-wide culture survey – January 1995

The second culture survey was conducted by RB involving all employees present at the time of the original survey conducted two years previously in January 1993.

2.3.15 Operational review process – 16 January 1995

The eighteenth 'merged' Management Meeting agreed an operational review process with each department to focus on contribution to policies. Each review was to last a half day and cover projects, QIDW and actuals against targets.

2.3.16 Newsletter handover – 16 January 1995

It was decided to hand over the production of the *Newsletter* to a staff committee.

2.3.17 Management Development Plan – 16 January 1995

A Management Development Plan for the Management Team was agreed, arising from earlier training needs analysis, and proposed dates were discussed.

2.3.18 Second company-wide culture survey results produced – 27 January 1995

The second culture survey results were produced, showing significant improvement in most elements of company culture since the original survey in January 1993. Extracts from the second survey are given below.

Purpose of survey

The purpose of this survey is to examine employees' perceptions of the change which has occurred in organization culture within GSi Travel and Transportation (UK), two years into the installation of the TQM process.

The survey consisted of 21 parameters which were expressed on the survey instrument as a continuum of behaviours reflecting the presence or absence of an approach to work and organization consistent with a Total Quality Management philosophy. Survey participants were asked to give their views on behaviour, attitudes and organization in the company as a whole, rather than their own personal approach to the issue. The format was identical to the survey conducted in 1993 except that participants were asked to enter two scores for each parameter. The first response was to indicate the culture of the company as they perceive it at present and the second response was to indicate the perception of the culture as it was in 1993, when TQM was being introduced. The purpose of the two scores was to enable a third indicator to be produced, i.e. the participants' perception of the **degree of change** or movement in culture experienced during the two years concerned.

Summary of conclusions

Significant movement, both perceived and actual, has occurred on the relevant parameters referred to in clauses 2.2 and 2.3 of the 1993 survey report. Attitudes toward training and change in the degree of communication and its direction represent some of the greatest perceived movements in culture.

The degree of change perceived by technical support staff in questions 6 and 7 (direction of communication and openness of decision-making) was positive but small.

Reward systems are not seen as having changed significantly to reflect the emphasis on quality.

There has been a marked reduction in the numbers of staff who perceive the culture as being on the lower end of the scale, reflected in the fact that there are no parameters with clusters of over 20% of responses scoring 0 to 1. This again emphasizes the improvements seen in the openness of decision-making (Q7), frequency of recognition (Q15) and leadership visibility (Q19), all of which had large clusters of negative perceptions in 1993.

Recognition and reward systems still require attention, despite some improvement in the latter. The advent of formal performance appraisal may affect this positively (although if handled badly, appraisal can actually produce the opposite result!) as may the development of better coaching skills amongst the Senior Management Team.

There appears to be a more cohesive view of the organization's culture than in 1993 with very few significant differences between employees' perceptions of the culture in different departments.

Key results for 1995 scores

The company has scored exceptionally well (an average of over 6) on its approach to quality, its customer orientation and the value it places on training. These results are reinforced by the fact that over 40% of employees scored them as 7 or 8.

Very few parameters had many scores in the range of 0 to 2, several having none at all.

The current set of results, compared to other surveys on the database, show that GSi has the best overall score on several parameters, notably those relating to attitude to quality, communications on business developments, planned decision-making, proactive management (low fire-fighting tendency), attitude to training and cooperation between departments.

In general, questions relating to employee motivation issues scored lowest, indicating a continuing need to further improve the involvement, recognition and reward processes as causal factors.

The effects of the culture as it now exists is likely to have beneficial effects on customer service, job performance and communications with small negative effects on employee attitudes. The last point is reinforced by the fact that question 20 (employee attitudes) recorded the second lowest perceived change in culture.

Differences in results at sub-group level

Very few important differences were observed when the data was analysed on a departmental or job basis. Differences were less pronounced by department than in the 1993 survey. The technical group again scored slightly lower overall but not significantly so except on one parameter:

- Effectiveness of communication on business developments (Q4)

However, there was still an improvement on their actual recorded 1993 score.

Differences between sub-groups in their perception of the degree of change in culture were relatively small. Other than the departmental analysis (see next point) the only score over 1 point adrift from the next nearest was for over-56 year olds on change in cooperation between departments. However, as this group consists of only six people, it is not considered to be important.

Technical staff scored the **degree of change** in culture significantly less than the other groups on the following questions:

- Attitudes toward risk taking (Q17)
- Employee attitudes to the company (Q20)

Change in culture, 1993 to 1995

Although the lowest scores still fall into the same categories as 1993 (openness of decision-making and reward systems), there is **perceived** to have been (but see below) positive movement on all parameters, with only these two scoring less than 4 on average.

The largest perceived movements (over 2 points) in culture relate to the company's approach to quality (Q3), the effectiveness of internal communications (Q4) and the value of training (Q18).

Least perceived movement has occurred on reward systems (Q14). In fact this is the only parameter which scored lower (fractionally) than in the 1993 survey. The data capture team seem to have perceived quite a large movement to have taken place but they in fact scored it more or less the same (5.1) for 1995 as they did in the 1993 survey.

People's perceptions of the change in culture, on most parameters represent about twice the degree of change actually measured by the results of the 1993 survey compared to this one.

When the 1994/5 results are compared against the actual scores from the 1993 survey, the gap is somewhat smaller on most parameters, but in all cases except reward systems it remains positive. The largest positive change in actual scores is in leadership behaviour (Q19) and the effectiveness of communication (Q4 and Q6).

The lowest actual positive change in scores relates to the use of motivation by fear (Q13), attitude to suppliers (Q2) and customer orientation (Q1). Since the last item was scored very high anyway in the original survey, the small degree of movement is not surprising.

Examined on a question by question basis, the survey shows that the number of people perceiving positive change to have occurred is several times the number who perceive no change or negative movement to have occurred. The largest groupings of no change or minus scores appeared on the following issues:

- Attitude to suppliers (Q2)
- Use of negative motivation (by fear) (Q13)
- Basis of rewards (Q14)
- Use of recognition (Q15)
- Attitude toward risk-taking and experimentation (Q17)
- Leadership behaviour (Q19)
- Employee attitudes (Q21)

In all the above cases, though, positive scores still predominated amongst respondents.

Recollections of members of the Steering Group/ Management Team of the implementation process

General comments

The Managing Director on middle managers:

There have been no signs of resistance from middle managers. The teams were picked by the senior management team. People with the relevant background who were asked to participate did not resist nor did people who were not asked complain. We communicated to everyone what we were doing and tried to make everything visible all the time. We displayed the templates around the building and encouraged people to look at them. We also encouraged managers to attend presentations. Once they got over the initial shock, they enjoyed it. Middle managers were encouraged to participate. For example, the DPA and QIDW were both cascaded down through the middle managers and we can't say the programme has stalled through resistance from them.

The TQM Training Manager:

The rest of the company is not as advanced as the Travel and Transportation (UK) part of the company (and we are not that far advanced). The rest of the seven activities are using MSI. Philippe Lesueur is trying to encourage TQM in all parts of the company. Our success is a result of the focus and drive of the Managing Director with support from Philippe.

There was not much success from the TQ process during the first year. Fast-track projects were poorly selected and the teams went through a period of low activity. This was mainly because the Steering Group assigned the team leaders but did not define the problems down to a low enough level. Not enough data was gathered about the problems beforehand. The Steering Group simply identified an issue as seen by the customer but did not take time to study it. Not enough support was given to the teams since there were no experts in the process. They took too long before they identified that they needed the support. Team members became demoralized and demotivated. MSI were then called in to provide further expertise and to help look at what was blocking them. Not all teams kept comprehensive minutes.

The Manager – Direct Marketing:

I am only involved in the Steering Group and not in any departmental Quality Circles or cross-functional teams. The whole process seemed slow, labour intensive and lacked real focus. There was a general distrust in the teams due to a lack of knowledge within the company. The problems chosen were too big. We tried to do too much at once.

I sponsored a Quality Circle last year. It completed its fourth and sixth templates this year. It has produced recommendations for improvements for one of our customers. There are tangible benefits for the customer rather than for GSi since the recommendations were about what the customer should be doing to help themselves. Other teams are now functioning better. We are still learning and now set shorter timescales and have a better awareness of the problem-solving process. There is a dramatic difference since we started in 1993. Since November 1994 there has been a noticeable change in the meetings. We discuss quality and business in the same breath – which is good.

There was too much talking – there was a lot of academic debate about what we should be doing. It was extremely time consuming. I resented the time taken away from more important jobs because we weren't focused. We suffered from a lack of knowledge and experience. There were one or two cynics in the Steering Group because so much emphasis was put on trying to resolve the underlying problems in the company. Now people have become less cynical because of the focus that we now have. They were cynical about the way we were introducing the TQ process, because of a lack of knowledge about the process. We've all changed.

The Operations Manager:

Implementation was poor because we were all tied up with our own particular areas. We discussed things a lot but were not always held accountable for actions. Now we have Quality in Daily Work our general performance is measured and there is a plan that we will all be appraised. All of us are in the same boat basically, we are all under pressure. It's a wonderful environment, never boring, there's always something different – you can't plan what you are going to do that day.

The Accounts Manager – BA Executive Club and BSP:
The TQM Training Manager looks after the groups now. He looks after all of us really well. If you've got a problem, you go to him. If the groups get stuck they go to him, he's always helpful.

General comments from staff

'I haven't thought about recognition. We went out for a meal. The company gave us a contribution towards the cost but it had to be prompted.'

'We also got mentioned in the *Newsletter*. It was definitely a nice feeling and it increased our motivation – we actually offered to come in in our own time to do things.

'I've been on an appraisals course and a winning edge – confidence course. It made me feel a lot better.'

'The courses have benefited me. They are good. It helps you look at yourself and at other people. The downside is that it's a while before you can put it into practice – it takes quite a long time. This can drop your morale.'

'We would like to see management have a better understanding of TQ. They made themselves available to us, eventually.'

'If new Quality Circles are starting up the members need to know there is expertise available.'

'It's confusing when different managers have different views about what is the best way to tackle a project; we're always so busy, not everyone can get involved because it will affect productivity.'

'We've got a lot of new members in the department, about 50% are involved. A lot have received training and there is another MSI course coming up.'

'All groups have sponsors – either a line manager or cross-functional person who raised the problem. We have a complaints procedure – they go through a funnel or system. On top of that AG and LG have facilitator roles.'

Observations of members of the Steering Group/Management Team on evidence of change benefits

The Managing Director:

- There is a much greater awareness throughout the company of what we are trying to do.
- There is a much greater lateral communication system than there was before.
- Certainly there is a greater realization of the impact of non-quality on our customers.
- There is a willingness to get things right or put things right.
- There is also a willingness amongst employees to get things right.
- The previous culture was top down driven. Now people are beginning to take the initiative themselves and realize they can have an impact.
- We are not that far advanced at the moment but think there is a significant change.
- Also there is a willingness to ask customers what they want.
- The style of management has changed. We are now more of a team. I am less dominant than I used to be – I hope.
- The management team is more focused – this is probably the most critical area.
- We now all work to the same agenda rather than individual agendas.
- We are more quality conscious.

The Manager – Direct Marketing:

- See the culture survey reports.
- There is a greater awareness of both internal and external customers.
- There is a greater awareness of the business by its employees.
- There is a greater awareness of the importance of profit and of maintaining and implementing quality.
- In spite of the current redundancies, the TQ programme should give greater job security to employees. In making the redundancies, we reiterated the quality approach and explained fully to everyone what was happening.
- There are still a few sceptics as a result of the problems with Quality Circles and the way they were run. Aspirations were built up but they didn't deliver in the early days.
- Initially about 30–40% of the people were committed to what we were doing. Now it's much higher – say, about 60–70%.

The Operations Manager:

- People are more open about admitting there have been problems.
- There is a greater awareness that doing things wrong costs somebody, somewhere money.
- There is strong support now for looking at costs of quality.

Training has broken down barriers between departments and even between functions within departments. As a result of exposure to tools and techniques we now use process flow charts. From these, people can see what they are supposed to do, how much they are actually doing and how silly some of the things that they do are. Now they want to know why they are doing it. It's still not part of everyday work – that's where QIDW will come into it. Work is not just about a series of problems – work is what the templates do for you.

The Data Preparation Manager:

The major benefits of the whole exercise are [pause] I don't know. Certainly more visibility for the Management Team. Also I've got to know the Management Team, which I didn't know before.

The Accounts Manager – BA Executive Club and BSP:

Quality Circles benefited the areas they were working in, especially the benefits of team work in general. We became very aware that we had a lot more people who could come up with ideas – people began to say 'Why didn't we do it like this before?' We became aware of people's skills because they had an opportunity to show them. We started to produced skills matrices. Quiet people came into their own. Some people, like myself, learned listening skills.

Staff talking about evidence of change and benefits

Evidence of change

'What happens now is that people are more aware of how it's meant to work. Before it was the Steering Group who decided. Managers are now experienced to take action. Managers are being proactive – taking more responsibility. They realize that the sooner they get things moving the less of a problem it is.'

'Early on projects took up to eighteen months, now problems can be handled in a much shorter time. There is an awareness that if we handle these things as we come up against them it's better for all concerned. There are still lots of things we haven't tackled but we are in a much better shape.'

'The integration of the Danbury approach and MSI gave us a headstart. We started MSI training in mid 1993 – we were already well aware of Pareto and histograms – it was a great reminder – two people were preaching the same story.'

'Richard Barnes concentrated on fundamentals – the focus was on proven and tested approaches. At the TQ open day most GSi companies appeared to be all over the place. We had a clearer idea of what we were trying to achieve. Richard Barnes kept our feet on the ground. QIDW was similar but came after.'

Benefits

- Loads of training for a more skilled bunch of people.
- Now have about 60% of people participating in how we manage the company. Before it was about 5%.
- People are asking more questions – is this the best way?
- We are constantly trying to do it better to get it right. We probably don't get it right every time but we are trying.
- I'm not part of the senior team – now its easier because I don't have to devote Monday afternoons to long, strenuous, nasty meetings!
- We are more commercially aware now. We recognize our customers should like us for the right reasons, we are much more customer focused than before.
- We've got to know the management team more.
- It helped me to know who everyone is.
- It's a lot more open and it has helped to break down barriers – some people might disagree with that.
- It has improved working between departments. It's nice to understand what goes on elsewhere.
- It's made us aware that we all have internal customers. It's difficult to see them as customers so we don't know what their needs are.
- My internal customer is the rest of the team. I need to communicate with them.

The biggest barrier

'It's the boss's job'.

'The way in which the senior team didn't gel. Now it's better – they do now.'

What would you have done differently?

The Managing Director:

It's difficult to say whether or not if anything was done differently it would have had a better impact. I think we would have done the projects differently – we would have chosen them differently and would have given more coaching to the team members.

The main area of disappointment is that projects took so long.

We would also have linked the process more closely with the business instead of treating them as two separate entities. We should have done that earlier.

The Operations Manager:

I would have tried to have done it – to put the ideas into action. We really needed to have the business going on whilst we were off doing TQ instead of trying to do our daily job at the same time.

Maybe it was a case of prioritizing things. In hindsight I can see a lot of opportunities that were missed – for example, getting the training and development plan in place at the outset using a company-wide approach rather than a departmental approach. QIDW should have been done earlier – you don't have to be in a team to use it.

The Manager – Direct Marketing:

It's taken a long time with no significant tangible benefits as yet. We will see these hopefully this year. It's been a painful process but I have no regrets. It's right for the company.

Conclusions

- Cultural change requires a very heavy investment of managers' time. The adoption of TQM in this company has been relatively widespread and in depth. Each of the top management team invested between 20 and 100% of their time to the process either in meetings of the Steering Group or in support to others or in carrying through actions associated with their personal culture change roles.
- The change process must have the wholehearted support of all the team members if it is to succeed.
- Stability within the team helps to sustain progress but lack of new blood can lead to stagnation of thinking.
- The role of the TQM Champion should not be underestimated.
- The lack of full understanding of the culture change process and what is required to drive it can hold up progress for a long time. In the absence of this full understanding, there is a tendency for people to fall back on doing the easy or mechanical things rather than address the more profound changes in thinking and behaviour which are required to achieve a TQM culture.
- The bottom-up process (setting up teams and projects) is easier to do than the top-down (Mission, CSF and policy deployment), but both must be in place if the project teams and the techniques they adopt are to be seen as more than just a fad. Management must also behave as they expect their staff to do in regard to tackling problems. That is, they must use the same processes for which they are training their teams, and be seen so to do! Leadership by example, in other words.
- Measurement is of paramount importance. In its absence, progress can be patchy and even fall away at times.
- Recognition and reward is an integral part of the process. Companies tend to skim over this aspect too superficially and fail to really challenge the basic premises of existing reward and recognition systems.
- Other forms of management training may be necessary to support the culture change process, such as the development of skills in coaching and project management.

Recollections of the consultant (R Barnes)

I remember the planning and strategy workshops being somewhat fraught at times. Unusually, there was a lot of competition between members of the Steering Group for particular culture change roles. In most organizations, members accept any role but I recall I got some criticism for the way I had handled the allocation of roles. It was felt at the end of the day to have been not very democratic and we got a few angry silences from members of the group.

There was, as is normal with such a change process, a fair degree of misunderstanding as to what each role was to achieve and they needed an extra day to review the relationships between the roles with me. Most of the group found this review day helped them slot the issues together.

The culture surveys speak for themselves. The group did make significant progress in opening up communications and involvement in the company but struggled with the strategic and measurement sides of the process. In the end they abandoned the concept of critical success factors in favour of a more traditional set of objectives based on four key policies. A framework was then developed to manage these but the team is still struggling with acceptance of the planning and target-setting aspects.

I was always very impressed with the personal level of commitment and time put in by several members of the Steering Group – their involvement in delivering significant chunks of management training and the competition to take on the culture change roles was indicative of this. They were never afraid to experiment and try out new approaches. At the end of the day though I am not sure that that level of commitment has spread significantly widely. Although there are lots of people involved I would have expected more champions of TQM to have emerged elsewhere among the staff by now than seems to be the case.

At times the group seemed intent on analysing things to death rather than doing things and there was always a reluctance to give those charged with particular tasks their head without wanting to alter and refine things to too great a level of detail. Anything can be made complicated if you work at it.

The management development programme which was put together aims to help the team develop some of the important management skills which are required to continue progress.

The role of MSI was very helpful. The group were able to draw on two sources of advice and split the project team process away from the strategic issues. However, I have reservations about the continued insistence on one single form of problem-solving and think the group should now be widening the education process to get people to appreciate the process of problem-solving rather than the mechanics of completing templates which can have a habit of becoming an end in themselves.

Glossary and references

Professor Kano and Dr Kondo	Japanese consultants employed by GSi at corporate level to oversee the implementation of TQM strategically across all countries.
MSI	Management Systems International – a Florida-based TQM consultancy with experience of TQM implementation in Florida Light and Power, a Deming prize winner.
ITO1	'Improvement Technology' training course designed by MSI – covers the first three templates within the problem-solving framework developed by MSI (see appendix for examples of templates used in stationery project).
ITO2	As ITO1 but covering the final three templates in the framework (see appendix for example of templates used in stationery project).
CSF	Critical success factor.
DPA	Departmental purpose analysis
UKD	UK director
DMD	Divisional managing director

Appendix to the case study

A1.1 PROPOSALS FROM DANBURY TOTAL QUALITY AND INNOVATION MANAGEMENT CENTRE FOR A TQM INSTALLATION PROCESS

A1.1.1 Overall aims and objectives of the project

Overall aim

To assist the senior management team within the company's Camberley offices to effectively determine a strategy for, and implement, a Total Quality Management process within the company.

Overall objectives

These are:

- Involvement of the senior management team of GSi Travel and Transportation in the preparation of total quality installation plans and their commitment to the implementation of actions required
- Creation of appropriate organization structures to manage the improvement process
- Involvement of employees at all levels in the organization in the process of continuous improvement
- Increase external and internal customer satisfaction
- Improved communications throughout the organization
- Continuous quality improvement and defect prevention
- Reduction of waste
- Elimination of non-added tasks
- Education and training of all employees to ensure the process will work

Basic approach – an overview

It is becoming increasingly recognized that the adoption of Total Quality principles can dramatically transform operational performance of almost any type of business or organization. Total Quality can lead to a competitive edge, something for which all organizations should be striving.

At first, the Japanese, because of the situation they found themselves in during the 1950s, adopted and enlarged the concept and achieved remarkable results. This unfortunately led to the claim that it depended on Japanese culture, and because of this the technique would not work elsewhere.

An increasing number of Western organizations have successfully achieved culture change through the adoption of Total Quality concepts and practices. They include IBM, ICL, Xerox, Caterpillar, Hewlett Packard and Rover. They have all achieved remarkable improvements which have led to their winning edge.

A host of others, attracted by the claims of endless improvements, have been unsuccessful. There is not one root cause which is common to most of them. They found it impossible to achieve a lasting change in the way that their employees, and especially managers, behave. A Total Quality culture cannot be installed without such a change, since the basic driving force of 'Continuous Improvement' will be absent.

These organizations correctly concentrated on 'Management Commitment', the most important element of a Total Quality culture. The managers concentrated on steering the implementation process; what they did not do is to apply their commitment to changing their own behaviour.

The approach which the Total Quality and Innovation Management Centre of the Anglia Business School takes to the installation of a Total Quality Process is directed at overcoming this problem. We direct our attentions to guiding and assisting management in the crucial role that they must play. We act as agents and facilitators of the radical changes which management must bring about, if their company is to be counted amongst those who have successfully installed a Total Quality Process.

This approach concentrates on the identification and speedy removal of barriers to business success, whilst building the 'bottom-up' approach which is essential to getting all employees involved in continuous improvement. In this way the Total Quality Process becomes management-led and provides both tangible 'bottom-line' benefits in the short term, and lasting improvements in the long term which come from a major change in company culture.

The need for a dual-track approach

When a Total Quality Process is successfully embedded in the culture of a business it delivers customer satisfaction in all activities, both within the inner workings of the organization and in its relationships with its external customers. To introduce a Total Quality culture can take four to five years in any business. Most businesses, however, need to provide tangible results to sustain top management support and gain commitment from all employees in the short term.

To address this paradox, we recommend a dual-track approach to installing a Total Quality Process at GSi Travel and Transportation. On the one hand we suggest the establishment of a fast track to achieve early success and fast results, whilst on the other hand we think there is a need for a culture change track in order to achieve a culture of improvement throughout the organization.

The fast track

The fast track should provide the foundation for the culture change track. It requires management leadership; the early identification of critical success factors for the business and of major opportunities for improvement related to these; and the preparation of priority action plans aimed at the need to remove barriers to success which may be present in the major business processes. The approach would be implemented by cross-functional improvement teams overseen by a Steering Group.

The culture change track

Building on the early successes of the cross-functional teams, the culture change track would be delivered through company-wide communication and training and through a company-wide education process. Part of this culture change process would be to create a climate in which continuous improvement would become the norm. It might require changes in leadership style, a willingness and ability on the part of management to delegate more effectively, improved decision-making processes and efforts to enlarge the roles of people lower down in the organization. The Total Quality Management process is also about changing employees' attitudes towards Quality which may require empowering them to exercise newly acquired roles and responsibilities. This in turn may require changes to organizational systems and policies.

To assist the company to identify the cultural barriers and drivers and the changes required, we recommend the use of a company-wide survey of all employees' perceptions of the company culture.

A1.1.2 Proposed programme of support

An important element of our approach is the establishment of a joint working relationship with you. A key measure of our success will be the extent to which ownership and further development of the Total Quality Process is successfully transferred to you once it is up and running. We see this process involving a number of phases.

A1.1.3 Phase 1 – Awareness and implementation planning

Top team two-day workshop – objectives

- A review of TQM philosophy and concepts
- Agreement of what quality means for GSi Travel and Transportation
- A review by senior management of the company's Quality Policy and Mission
- A review of the business to identify factors critical to its success
- The identification of obstacles to the installation of a Total Quality Management process and the preparation and agreement of plans for the removal of these
- The preparation of a draft Total Quality Management installation plan including proposals for the training of managers and other employees, taking account of existing or planned competency-based training provisions
- The identification of the key barriers to business success and the establishment of self-managing teams, where appropriate, to create improvements
- The identification of important measures of performance
- Agreement of 'fast-track' projects for self-managing teams
- The establishment of a TQ Steering Committee to guide the project
- A Communication Plan

A1.2 REVISED PROPOSALS FROM DANBURY TQM CENTRE

A1.2.1 Phase 2 – Awareness and fast-track project team training

Middle management awareness

Because line managers are critical to the success of the process and are also often the group with potentially the greatest changes facing them, phase 2 includes a two-day awareness seminar for middle management, supervisors and key professional staff.

This workshop will deal with both the strategic issues behind Total Quality Management and the implementation process, including their role in it. A programme for this course is attached (not reproduced here).

Company-wide awareness and fast-track project team training

Company-wide awareness will be achieved through a series of in-house presentations. Training in Total Quality Management tools and techniques will be provided for groups chosen to undertake fast-track projects. This will be provided by MSI trained trainers.

This phase will also include implementation support to the Steering Group, including a training needs analysis and a team-building programme.

Consultant time

- Management awareness seminar: 2 x two-day courses, plus 1 day preparation time = 5 consultant days
- Support to Steering Group including training needs analysis and team building = 20 days (estimated)
- Total = 25 consultant days

A1.2.2 Phase 3 – Managing the improvement process

Facilitator/team leader training and tools and techniques training

We understand these requirements will be met through MSI.

Departmental purpose analysis

It is recommended that the company adopt a structured approach to the identification and development of the internal customer/supplier chain through the use of a training package known as PAT (Purpose, Assessment, Targets). Details of the package have been provided separately.

Consultant support estimated for the use of this pack is as follows. The content of support in Phase 3 will depend heavily on your own resources and perceived needs. However, we would estimate the need for consultant help in some of the Steering Committee tasks such as departmental purpose analysis, project support, etc. Quarterly implementation reviews with the Steering Committee over the first two years of the process are also recommended.

Consultant time

- Estimated number of consultant days assuming the use of PAT pack for three divisions = 10 days
- Steering Committee support = 4 days in year two (estimated)
- Total = 14 days

A1.3 EXTRACTS FROM A TQM NEWSLETTER

Quality Circles roll out in Clearing

This month has seen a flurry of activity in the Clearing Department with the formation of four Quality Circles.

CRS

Tina Jackson and her team will be looking at improving the CRS process across Cash and Credits.

Fred's area

Sue Tanner and her evening shift team will be addressing the age-old problem of getting on to the main sortbox area (referred to as Fred by those in the know).

Edit

Margaret Jones and her edit contingent are looking at the time currently spent on the edit function: this keen team has already had two meetings. Well done!

AVS

Ann Bush and her team of supervisors will be addressing the area of Airline Validation Services.

Prior to their first meetings, the individual teams each attended a very enthusiastic presentation on Quality Circles given by Jane Leverton. Jane gave a brief overview of Total Quality and the role of Quality Circles, where to go for help etc., in addition to a few hints on meeting rules etc. From all those who attended... Thanks, Jane.

If you would like any information on the Circles, please contact the team leaders concerned or myself (Lesley Goodall) on Ext. 2352

QUALITY CIRCLE *DIRECT MARKETING*

Improving correspondence turn around time

SUSPECTED ROOT CAUSES SELECTED

At the time of the previous newsletter going to press, we had completed Template 1. Since then, we have progressed through the templates and are currently tackling Template 3.

It has not always been a straight forward route through. After Katy Gillibrand's excellent presentation to the Paris TQ Conference, it was decided that we needed to re-address our indicator in Template 1. This has been amended to "Files awaiting action outside of a 10 day turnaround". The data display has been amended to reflect this.

Template 2 was completed fairly quickly as we all work within the same department and are familiar with the process shown on our flowchart.

Template 3, however was another matter. Once we started expressing our ideas on possible root causes, we could not stop and ended up with more fishbones than HM The Queen Mother.

We managed to select four suspected root causes from the fishbone diagram: Staff motivation; Staff training; Usage of office equipment and No memory to re-printer letters. We are currently collecting data which will enable us to display data to back our findings so far. We are confident that we have selected valid root causes but that is a risky statement at this stage of the game!

To find the results, please read the next update.

ALAN BUSH (Team Leader)

QUALITY CIRCLE *SYSTEMS & PROGRAMMING*

Improve system software control procedures

PUTTING THEORY INTO PRACTICE

After returning from an enjoyable IT01 (Improvement Technology, Part 1) course, a couple of Systems & Programming analysts decided to put theory into practice by setting up a Quality Circle to tackle a then unknown project. Three further S & P personnel were asked if they wished to participate, all of whom readily accepted.

For our first meeting, we agreed to have a brain-storming session to determine an appropriate project that we could initiate. After much deliberation, we chose the above project believing it to be beneficial to our customers who, for this project, would be our immediate colleagues.

The project, in a nutshell, is to review the current software control procedures and enhance them, if necessary, to cope with expanding business developments. Software control is the set of procedures we follow to create, store and maintain application programs and their corresponding documentation.

Having only had three meetings in total so far, you can forgive us for not redecorating the company's corridors with our findings as yet. In reality, we have completed the Background Information and are in the throws of circulating a questionnaire to establish the Customers' Viewpoint.

Our next step is to summarise the information received back from the questionnaires in order to complete section 2 of Template 1. More news in the next issue.

PETE KENYON (Team Leader)

The Editor's Letter

One of the key principles of Total Quality is continuous improvement and I hope you are seeing an improvement in the newsletter as we become more experienced in Total Quality.

Putting the newsletter together proves easier, and more enjoyable every time we do it.

Our learning points, so far, are:-

- Set a target date for completion/delivery;
- Plan each task well in advance and allow contingency;
- Let others, especially those who have been asked to contribute, have a copy of the plan;
- Remind contributors of impending publication dates;
- Encourage participation ("You WILL do an article for the newsletter, won't you?");
- More pictures (watch out for Kate with her old box Brownie);
- Never turn down offers of help.

Last, but not least, my sincere thanks to all those who have contributed to this bumper edition, especially to those who volunteered their services.

A Merry Christmas to you all. *LINDA WAKERLEY (Editor)*

WHAT DOES QUALITY COST?

A team has been formed to look at Cost of Quality issues.

Quality costing is the means of monitoring the effect of doing things wrong. It is a powerful technique for getting management attention, for prioritising problems and showing improvements.

The best way to measure Total Quality is to calculate the real effect to the business of failing to achieve requirements.

When a requirement is not achieved correctly first time, the full cost impact is classified as a COST OF NON CONFORMANCE (CONC).

Non conformances are thus measured in £££s, not in numbers, percentages etc. This focuses attention, enables prioritisation and leads to the justification of corrective action.

The experience of many companies is that focusing on the most urgent improvement areas is facilitated readily by the employment of Quality Cost Measurement.

The team will put in place procedures to gather information regarding the following:

Cost of Conformance (COC)

This can be defined as the cost of investing to ensure activities are carried out correctly the first time, every time and that problems are prevented.

Cost of Non Conformance (CONC)

This is the cost incurred by failing to get an activity right first time, on time, every time.

The team is at a learning stage at present but we hope to have our initial measurements in place during December.

RESPECT FOR PEOPLE

In 1992, the top management of GSi (Corporate) established five policies:

- Respect for people
- Focus on prevention
- Customer satisfaction
- Management by fact
- Continuous improvement

For those of you who have attended recent GSi training courses, these policies (or core values as they are often referred to), may already be familiar to you, but for those of you who have not been exposed to them, it is proposed to take each one in turn. The spotlight for this edition is "Respect for people".

"Respect for people" is again more commonly referred to as the "Rules of Life" (I hope you are following this because there will be questions later...) and is covered more fully on the Tillard Training Courses. Basically, the "Rules of Life" is the culture of the organisation: the way of life.

Throughout this edition, you will see symbols which denote one of the Rules of Life... but what do they mean? Only the text will tell! Turn to page 7.

APPPENDICES

A Useful addresses

British Deming Association
2 Castle Street
Salisbury Wilts SP1 1BB
Tel: 01722 412138
Fax: 01722 331313

National Society of Quality Circles (NSQC)
2 Castle Street
Salisbury
Wiltshire SP1 1BB
Tel: 01722 26667

European Federation of Quality Circles Associations (EFQCA)
44 Rue Washington
Brussels 1050
Belgium

Institute of Quality Assurance
PO Box 712
61 Southwark Street
London SE1 1SB
Tel: 0171 401 7227
Fax: 0171 401 2725

British Quality Foundation
Vigilant House
120 Wilton Road
Victoria
London SW1V 1JZ
Tel: 0171 931 0607
Fax: 0171 233 7034

The National Quality Information Centre (NQIC)
PO Box 712
61 Southwark Street
London SE1 1SB
Tel: 0171 401 7227

The National Accreditation Council for Certification Bodies (NACCB)
2nd Floor
3 Birdcage Walk
London SW1H 9JH
Tel: 0171 222 5374

The TQM Magazine
James Creelman
MCB University Press
60/62 Toller Lane
Bradford
W. Yorks BD8 9BY
Tel: 01724 499821
Fax: 01724 547143

Quality Matters – Newsletter
Spire City Publishing
PO Box 81
Abingdon
OX14 2PJ
Tel: 0836 520665

European Foundation for Quality Management
Building Reaal
Fellenoors 47a
5612 AA Eindhoven
Holland
Tel: +3140 461075
Fax: +3140 432005

The Bristol Quality Centre
PO Box 54
Fishponds
Bristol BS16 1XG
Tel: 0117 9656261
Fax: 0117 9583757

The Total Quality & Innovation Management Centre
Anglia Business School
Anglia Polytechnic University
Danbury Park Conference Centre
Danbury
Essex CM3 4AT
Tel: 01245 222141
Fax: 01245 224331

British Standards Institution (BSI)
Enquiry Section
Linford Wood
Milton Keynes MK14 6LE
Tel: 01908 221166

Pera International
Nottingham Road
Melton Mowbray
Leicestershire LE13 0PB
Tel: 01664 501501
Fax: 01664 501264

B Glossary of terms

Assignable cause. Any departure, or variation, from the plan, which was not accounted for in the process when it was set up. Assignable causes must be removed before common causes can be tackled in a process improvement task.

Attribute. A measurement of a factor for statistical control which is described as a yes/no binary system of data collection: e.g. the delivery is either on time or late. Scaling is not applied.

Bar X. The average of a series of observations of a process or product.

Brainstorming. A way of collecting ideas from a group of people.

Benchmarking. A process of identifying improvement targets by examination of cross-industry, best-in-class performance within a particular process.

CAR. Corrective Action Request. A form used to highlight and seek action on any non-conformity in the operation.

CARP. Corrective Action Review Panel. A subgroup of the steering committee directing CARs.

C-chart. A control chart using a count of the number of defective items in a sample.

Cause and effect diagram. See fishbone technique

Common cause. A cause of variation inherent in the process as it was originally set up. Common causes result in random variation from the target or plan.

Control chart. A chart with upper and/or lower control limits which are plotted values of some statistical measurements for a series of samples or subgroups.

Control limits. Limits on a control chart that are used as criteria for action, or for judging whether a set of data does or does not indicate lack of control.

Correlation. See scatter plot.

Cost of Quality. The combination of costs of ensuring conformance to specification or requirements, together with failure costs; sometimes shown as costs of prevention, inspection, internal failure and external failure.

Customer. Anyone who receives the results (output) of your work. Customers can be external or internal (colleagues).

Defect. Any non-conformance of an item to specified requirements.

DIG. Department Improvement Group. The members of any department working on improvements to their own processes and outputs.

DPA. Departmental Purpose Analysis. A method of identifying the work processes, inputs and outputs of a department.

Error-cause removal. A formal system for employee involvement in problem identification.

Fishbone technique. An analytical tool used in problem solving to examine potential cause and effect relationships.

Fixing. Putting something right without finding the root cause, thus allowing the problem to recur.

Hoshin kanri. See Quality Policy deployment.

How–how diagram. A technique for team-based idea building for maximizing problem solutions.

Inputs. Those things you receive from suppliers which enable you to do your job.

Ishikawa diagram. See fishbone technique.

JIT. A production control system for minimizing stocking levels and increasing throughput speeds.

Kanban. A system for materials control, utilized especially in JIT, which ensures minimum stocking, prompt renewal of stocks and delivery of the right materials to the right place at the right time.

Kaizen. The Japanese system of continuous improvement.

MRPII. Manufacturing Resource Planning. A system of production control incorporating manual and computer controlled systems to optimize materials and capacity utilization. The system is much more extensive than MRP – Materials Resource Planning, and is based on a strategic assessment of business needs prior to introduction, which is usually team-based.

NP-chart. A control chart showing the number of defective units in a sample.

Non-conformance The non-fulfilment of specified requirements.

Outputs. Those things produced by you for your customers.

Paired comparisons. A decision-making process producing a rank order of ideas by comparing each item in the group with each other item in turn and scoring 0/1 or 0/1/2 for each pair.

Pareto analysis. A way of defining the 'critical few' causes of problems or non-conformances.

Pareto principle. The 80/20 rule of thumb that in problem solving, (and other matters) 80% of the observed occurrences of a problem will be accounted for by 20% of potential causes.

P-chart. A control chart showing the percent of defective items in a sample.

PIG. A Process Improvement Group. A cross-functional team established on an ad-hoc basis to tackle a particular problem which concerns or is influenced by more than one department.

POC. The Price of Conformance. The cost of getting it right first time, made up of the costs of prevention and necessary inspection.

Poka-Yoke. A process designed by Shigeo Shingo to assist work groups to mistake-proof their processes.

PONC. The Price of Non-Conformance. The costs of getting it wrong.

Problem. A variation from the expected or required result.

Process. Any task in which some input is translated to an output through the actions of an individual, chemical/electrical reaction or machine.

Process capability. The limits of inherent variability within which a process operates.

Process mapping or flowcharting. A means of documenting the way a process works diagrammatically.

Process quality control. That part of quality control that is concerned with maintaining process variability within the required limits.

Quality. Conformance to the agreed requirements of the customer.

Quality Assurance. All the planned and systematic actions necessary to provide adequate confidence to management and customers that a product or service will satisfy given requirements for quality.

Quality Control. The operational techniques and activities that sustain the product or service quality to specified requirements.

Quality Manual. A document setting out the general quality policies, procedures and practices of an organization.

Quality Plan. A document setting out the specific quality practices, resources and sequence of activities relevant to a particular product, service, contract or project.

Quality Policy. The overall quality intentions and direction of an organization as regards quality, as formally expressed by top management.

Quality Policy Deployment. A formal process of objective setting derived from the Company Quality Policy and involving all departments and employee teams in planning improvements to support the corporate goals.

Quality Function Deployment. A technique for product or process design which marries together and optimizes the key quality requirements from the point of view of all interested parties, with a special emphasis on the end-user.

Quality System. The organizational structure, responsibilities, procedures, processes and resources for implementing quality management.

R-chart. A chart showing changes in the ranges of results observed.

Random variation. See common cause.

Range. The difference between the lowest and highest value of a variable.

Reliability. The ability of an item to perform a required function under stated conditions for a stated period of time.

Requirement. A description of an input or output which specifies the detail of how that input or output needs to be received.

Root cause. The original cause of a problem which, when eliminated, will result in no further occurrences.

Scatter plot or diagram. A tool for determining cause and effect relationships between two variables.

Simultaneous Engineering. An approach to new product development focusing on bringing the needs of the customer into the design process whilst involving all parties concerned with the eventual outcome into the design process as a team. The design team will normally include engineering, manufacturing, procurement, marketing and often involve suppliers. (See also Quality Function Deployment.)

SPC. Statistical Process Control. A method of measuring and improving the performance of a process by monitoring the results of the process statistically.

Special cause. See assignable cause.

Steering committee. The group responsible for directing and monitoring the quality improvement process.

Supplier. Anyone who provides you with material, information or a service in order to do your job. Suppliers can be external or internal (colleagues).

Taguchi methods. A range of techniques aimed at improving experiment design and the acceptance of parts to eliminate failures.

Variable. Any factor in a process which is not constant. Measured by some scale such as size, weight etc.

Why-why diagram. An alternative to the fishbone technique for analysing problem causality chains through a logical divergent thinking approach.

Index

Page numbers appearing in **bold** refer to figures

Printed in the United States
64926LVS00001B/107

9 780412 715303